SIGNIFICANT CHANGES TO THE

INTERNATIONAL
FIRE CODE

2006 EDITION

THOMSON
DELMAR LEARNING

Australia   Canada   Mexico   Singapore   Spain   United Kingdom   United States

THOMSON

DELMAR LEARNING

Significant Changes to the International Fire Code 2006 Edition
Jeffrey M. Shapiro, PE, FSFPE

Vice President, Technology Professional Business Unit:
Gregory L. Clayton

Product Development Manager:
Ed Francis

Editorial Assistant:
Jaclyn Ippolito

Director of Marketing:
Beth A. Lutz

Executive Marketing Manager:
Taryn Zlatin

Marketing Specialist:
Marissa Maiella

Director of Production:
Patty Stephan

Production Manager:
Andrew Crouth

Content Project Manager:
Kara A. DiCaterino

Art Director:
Robert Plante

Library of Congress Cataloging-in-Publication Data:
Card Number:

Application submitted.

ISBN-10: 1-4180-5301-5
ISBN-13: 978-1-4180-5301-7

NOTICE TO THE READER

Contents

PART 1
Administration and Definitions (Chapters 1 and 2)   1

- **102.3–102.5**
  Change of Use or Occupancy   2

- **Table 105.6.20**
  Permit Amounts for Hazardous Materials   4

- **106.4**
  Missed Violations–Approvals   6

- **202**
  Definitions for Emergency Shutoff Controls and Valves   8

- **202**
  Definition for Decorative Materials   10

- **202**
  Definition for Fail Safe   11

- **202**
  Definition for Assembly Group A   12

- **202**
  Definition for Business Group B   14

- **202**
  Definition for High-Hazard Group H, Exceptions   15

- **202**
  Definition for High-Hazard Group H-3   18

- **202**
  Definition for Residential Group R   20

- **202**
  Definition for Moderate-Hazard Storage, Group S-1   22

PART 2
General Safety Requirements (Chapters 3 and 4)   23

- **308.3.8**
  Open Flame Ignition Sources in Group R-2 Dormitories   24

- **309.5**
  Refueling of Powered Industrial Trucks and Equipment   25

- **311.5**
  Placards for Unsafe Buildings   26

- **313.1**
  Storage of Fueled Equipment in Buildings   28

- **404.2**
  Fire Safety and Evacuation Plans for Group B   30

- **404.2**
  Fire Safety and Evacuation Plans for Group R-2   32

- **Table 405.2**
  Frequency of Emergency Evacuation Drills   35

■ **408.3**
Emergency Planning and Preparedness for
Group E Occupancies and Group R-2 College
and University Buildings   38

PART 3
Building and Site Requirements   41
(Chapters 5–10)

■ **605.10**
Portable Electric Space Heaters   44

■ **606.8**
Refrigerant Leak Detectors   46

■ **606.9.1**
Remote Controls for Refrigeration Systems   47

■ **606.9.3**
Manual Emergency Control Boxes
for Refrigeration Systems   48

■ **606.10**
Emergency Pressure Control
for Refrigeration Systems   50

■ **608, 609, and 602.1**
Stationary Battery Systems   54

■ **Chapter 8**
Interior Finish, Decorative Materials,   59
and Furniture

■ **901.6.2**
Records for Fire Protection Systems   68

■ **901.9**
Recall of Fire Protection Components   70

■ **903**
Automatic Sprinkler Systems—   72
One- and Two-Family Dwellings and
Townhouses

■ **903.2.1.2**
Group A-2—Automatic Sprinkler Systems   74

■ **903.3.1.2.1**
Balconies and Decks—Automatic Fire   76
Sprinkler Installation Requirements

■ **904.11.5.1**
Portable Fire Extinguishers for Solid-Fuel   78
Cooking Appliances

■ **904.11.5.2**
Class K Portable Fire Extinguishers   80
for Deep-Fat Fryers

■ **905.3.7**
Standpipe Systems for Marinas   82
and Boatyards

■ **905.4**
Location of Class I Standpipe   84
Hose Connections at Horizontal Exits

■ **906.2**
Electronic Monitoring for Portable   86
Fire Extinguishers

■ **907.2.6**
Fire Alarm Systems for Group I   88

■ **907.2.9**
Manual Fire Alarm Boxes   91
in Group R-2

■ **907.2.12.2**
Emergency Voice/Alarm   92
Communication Systems

■ **907.10.1.2**
Alarm Notification Appliances   94
in Employee Work Areas

■ **907.15**
Monitoring of Fire Alarm   95
and Detection Systems

■ **910.1**
Smoke and Heat Vents for ESFR Sprinklers   97

■ **910.2.2**
Smoke and Heat Vents for Group H   99

■ **Table 910.3**
Smoke Venting and Draft Curtains
for High-Piled Storage   101

■ **914**
Fire Protection for Special Uses   104
and Occupancies

■ **1002.1**
Definition of Accessible Means of Egress   110

■ **1003.2**
Minimum Ceiling Height   111

■ **1003.3.2**
Projection Limits on Freestanding Objects   112

■ **1004.1**
Determination of Design Occupant Load   114

■ **Table 1004.1.1**
Occupant Load Determination for Day   116
Care Uses

■ **1004.2**
Maximum Occupant Load Permitted   118

■ **1004.7**
Occupant Load Determination
for Fixed Seating   120

■ **1007.1**
Platform Lifts as Accessible Means
of Egress   122

■ **1007.3, 1007.4, 1007.6.2**
Required Areas of Refuge   123

■ **1008.1.1**
Minimum Door Width
in Group R-1 Occupancies   126

■ **1008.1.2**
Door Swing in Sleeping Units   128

■ **1008.1.6**
Thresholds at Residential Exterior Doors   130

■ **1008.1.8.7**
Remote Unlocking of Stairway Doors   132

■ **1008.1.9**
Panic and Fire Exit Hardware   134

■ **1009.3.1, 1009.7**
Curved Stairways   136

■ **1009.5.2, 1010.7.2, 1014.5**
Weather Protection of Exterior
Egress Components   138

■ **1009.5.3**
Enclosed Usable Space under Stairways   140

■ **1009.10, 1010.8, 1012**
Handrails for Stairways and Ramps   142

■ **1009.11.2, 1013.5, 1013.6**
Protection at Roof-Hatch Openings   145

■ **1010.6.3**
Minimum Ramp Length   147

■ **1010.9, 1010.9.1, 1010.9.2**
Edge Protection at Ramps   149

■ **1013.3**
Guard Opening Limitations
for Group R-2 Occupancies   151

■ **1014.2**
Egress through Intervening Spaces   153

■ **1014.2.1**
Egress through Adjoining Tenant Spaces   156

■ **1014.3**
Common Path of Egress Travel
in Group R-2 Occupancies   158

■ **1014.4.2**
Aisle Accessways in Group M
Occupancies   160

■ **Table 1015.1**
Single Means of Egress from
Day Care Uses   162

■ **1015.2.2**
Egress Separation of Three or More Exits   164

■ **1020.1**
Unenclosed Interior Exit Stairways   166

■ **1020.1.7.1**
Egress from Smokeproof Enclosures   168

■ **1025.3**
Egress from Group A Occupancies   170

■ **1028.2**
Reliability—Maintenance of the Means   171
of Egress

■ **1028.4**
Exit Signs—Maintenance   172

■ **1028.7**
Testing and Maintenance—   173
Communication Systems for Areas
of Refuge

PART 4
Special Processes and Uses   174
(Chapters 11–26)

■ **Chapter 15**
Flammable Finishes   176

■ **1504.2**
Location of Spray-Finishing Operations   182

■ **1504.7.6**
Ventilation Termination Point   184

■ **Table 1805.2.2 and Section 1805.2.3.5**
Maximum Quantities of Hazardous   185
Production Materials (HPMs) at a
Workstation

■ **2205.6**
Warning Signs   189

■ **Table 2206.2.3**
Minimum Separation Requirements
for Aboveground Tanks   191

■ **2209.3.2 and 2211**
Location of Dispensing Operations
and Equipment for Hydrogen Motor Fuel   193

■ **2308.2.1**
Plastic Pallets and Shelves in Rack Storage   199

PART 5
Hazardous Materials (Chapters 27–44)   202

■ **Section 2701.1, Exception 10**
Hazardous Materials—Scope   204

■ **Table 2703.1.1(1) and Section 2702.1**
Maximum Allowable Quantities per   205
Control Area for Explosives

■ **Table 2703.1.1(1)**
Maximum Allowable Quantities per   207
Control Area for Combustible Fibers

■ **Table 2703.1.1(1)**
Maximum Allowable Quantities per   209
Control Area for Fuel in Fuel Tanks
and Piping Systems

■ **2703.2.9**
Testing of Hazardous Materials   211
Equipment, Devices, and Systems

■ **Table 2703.8.3.2**
Design and Number of Control Areas   214

■ **2703.8.3.4**
Fire-Resistance Rating Requirements   216
for Control Areas

■ **Chapter 28**
Aerosols   218

■ **Section 2902.1**
Definitions for Combustible Fibers   223

■ **3003.3**
Pressure Relief for Compressed Gas   225
Containers, Cylinders, and Tanks

■ **3003.16**
Vaults for Compressed Gases   227

■ **Chapter 33**
Definition of Explosives   231

■ **Chapter 33**
Separation Distances for Explosives   233

■ **Chapter 33**
Separation Distances for Explosives—   238
Day Boxes and Operating Buildings

■ **3308.2**
Permit Application for Fireworks   240

■ **3308.9**
Post–Fireworks Display Inspection   241

■ **3402.1**
Definition of Liquid Storage Warehouse   242

■ **3404.2.7.5.8 and 3406.4.6**
Overfill Prevention   244

■ **3404.3.1**
Design, Construction, and Capacity of   246
Containers and Portable Tanks

■ **3404.3.2.3**
Number of Storage Cabinets   249

■ **3404.3.5.1**
Basement Storage   250

■ **Table 3404.3.6.3(2)**
Storage of Unsaturated Polyester Resin   252

■ **Table 3404.3.6.3(3)**
Quantity Limits for Liquid Storage   255
Warehouses

■ **3405.3.8.4**
Weather Protection   256

■ **3405.5**
Alcohol-Based Hand Rubs Classified as   258
Class I or II Liquids

■ **3504.2.1**
Distance Limit to Exposures for   261
Flammable Gases

■ **3702.1, 3704.2.2.10 and 1803.13**
Physiological Warning Thresholds   263

■ **3704.2.2.7**
Treatment Systems for Toxic Gases   265

■ **3806.2**
Overfill Prevention for LP-Gas Containers   267

■ **3809.12**
LP-Gas Storage Outside of Buildings   269

PART 6
Reference Standards (Chapter 45)   274

PART 7
Appendices (Appendices A–G)   275

■ **Appendix B105.2**
Fire-Flow Requirements for Buildings   276
Other Than One- and Two-Family
Dwellings

Index   279

Preface

The International Code Council (ICC), publisher of the *International Fire Code®* (IFC), was established in 1994 as a not-for-profit organization dedicated to developing, maintaining, and supporting a single comprehensive, coordinated set of national model building and construction codes. Its mission is to provide the highest quality codes, standards, products, and services for all concerned with the safety and performance of the built environment.

The IFC is 1 of 14 International Codes® published by the ICC. This comprehensive fire code establishes minimum regulations for firefighter safety, fire protection systems, and the safe storage and use of hazardous materials using prescriptive and performance-based provisions. It is founded on broad-based principles that make possible the use of new materials and new building designs. The IFC is available for adoption and use by jurisdictions. Its use within a governmental jurisdiction is intended to be accomplished through adoption by reference, in accordance with proceedings establishing the jurisdiction's laws.

Many believe that the IFC is only a building maintenance document. This assumption is incorrect. The IFC is also a design document because one cannot construct or erect any building without fire department access, a water supply, and the required fire protection systems. Add a special hazard such as a refrigeration system, stationary lead acid battery system, or a diesel fuel or liquefied petroleum gas–supplied emergency generator, and additional design requirements in the IFC will be invoked.

The purpose of *Significant Changes to the International Fire Code 2006 Edition* is to familiarize fire officials, building officials, plans examiners, fire inspectors, design professionals, and others with the many important changes in the 2006 IFC. This publication is designed to assist those code users in identifying the specific code changes that have occurred and, more important, understanding the reason behind the changes. It is also a valuable resource for jurisdictions in their code adoption processes, explaining the technical changes associated with updating from an older edition of the IFC.

This book is arranged to follow the general layout of the IFC, including its code sections and section number format. The table of contents, in addition to providing guidance in the use of this publication, allows for quick identification of the significant code changes that occurred in the 2006 IFC.

Throughout the book, each change is accompanied by a photograph, an application example, or an illustration to assist and enhance the reader's understanding of the specific change. A summary, discussion of significance, and source code change proposal number for each of the changes are also provided (code change proposal numbers can be used as a tool in researching additional history on any change in ICC's "Code Change Resource Collection–2006" books, available from the International Code Council). Each code change is identified by type, be it an addition, modification, clarification, or deletion.

Each code change discussed in this book includes a mark-up of the actual code text, presented in the same format used for code-change proposals. Deleted text is shown with a strike-through and new code text is identified by underlining. This book presents the 2006 code text as compared with the 2003 text, so the user can easily identify changes.

As with any code-related text, *Significant Changes to the International Fire Code 2006 Edition* is best used as a study companion to the 2006 IFC. Because only a limited discussion of each code change is provided, the IFC itself should be referenced in order to gain a more comprehensive understanding of the code and its application.

The commentary and opinions set forth in this text are those of the author and do not necessarily represent the official position of the ICC. In addition, they may not represent the opinions of any enforcing agency, and such agencies have the sole authority to render interpretations of the IFC.

Acknowledgments

Many thanks to all who assisted in reviewing and supplying photographs for this book. Contributors of photographs or illustrations are individually acknowledged where these appear in the text. To ICC staff member Scott Stookey, I extend great appreciation for your unique contributions in editing and reviewing the text and for providing numerous illustrations. To Emily, Sara, and Michael, thank you for sacrificing our time together for what turned out to be a much bigger job than I thought it would be.

About the Author

Jeffrey M. Shapiro, PE, FSFPE, is President of International Code Consultants, a consulting fire- and life-safety engineering firm based in Austin, Texas, and is recognized as one of the nation's leading fire code experts. Mr. Shapiro holds a Bachelor of Science degree in Fire Protection Engineering from the University of Maryland, and he is a Registered Professional Engineer. Mr. Shapiro is a Fellow and past chapter president of the Society of Fire Protection Engineers, and he is a member of the International Code Council, the National Fire Protection Association (NFPA), Tau Beta Pi National Engineering Honor Society, and Salamander Fire Protection Engineering Honor Society. He has also served as adjunct faculty for the United States National Fire Academy.

Mr. Shapiro previously served as chief executive of the International Fire Code Institute (IFCI), a legacy partner with the International Code Council. Under Mr. Shapiro's direction, the institute became a leading fire-safety organization. The most well-known function of the IFCI was its role in administration of the Uniform Fire Code, which, at the time, was the fire code used in 25 states and by several agencies of the United States government. Mr. Shapiro also served as creator and executive editor of the *IFCI Fire Code Journal* magazine.

Prior to serving with the IFCI, Mr. Shapiro was Coordinator of the Uniform Fire Code for the International Conference of Building Officials and the Western Fire Chiefs Association. In this position, Mr. Shapiro served as the chief engineer responsible for overseeing development of the Uniform Fire Code, issuing code interpretations, developing and conducting professional development seminars, and assisting with development and application of fire- and life-safety provisions in the Uniform Building Code.

Mr. Shapiro has also previously served as Manager of Fire- and Life-Safety for Embassy Construction for the United States Department of State on a $2.9 billion project to construct new embassies around the world; Major Fires Investigator for the United States Fire Administration; and Assistant to the Fire Marshal for the City of Fort Worth, Texas, Fire Department.

SIGNIFICANT CHANGES TO THE
INTERNATIONAL
FIRE CODE

2006 EDITION

PART 1

Administration and Definitions

Chapters 1 and 2

■ **Chapter 1**   Administration
■ **Chapter 2**   Definitions

The provisions of Chapter 1 establish the requirements for the application, enforcement and administration of the requirements in the *International Fire Code.* In addition to establishing the scope of the Fire Code, the chapter addresses construction, design, operational and maintenance provisions, the adoption of certain standards that are found in Chapter 45, and the general authority and responsibilities of the Fire Prevention Division or Bureau. The Code itself is not enforceable unless it is adopted by a jurisdiction. Once adopted, Chapter 1 specifies the requirements for issuing and enforcing construction and operational permits, establishing a board of appeals, inspections of sites, buildings and processes, and actions required to serve and correct violations. Chapter 2 provides definitions for terms used throughout the *International Fire Code* and the *International Building Code.* ■

102.3–102.5

Change of Use or Occupancy

TABLE 105.6.20

Permit Amounts for Hazardous Materials

106.4

Missed Violations–Approvals

202

Definitions for Emergency Shutoff Controls and Valves

202

Definition for Decorative Materials

202

Definition for Fail Safe

202

Definition for Assembly Group A

202

Definition for Business Group B

202

Definition for High-Hazard Group H, Exceptions

202

Definition for High-Hazard Group H-3

202

Definition for Residential Group R

202

Definition for Moderate-Hazard Storage, Group S-1

102.3–102.5

Change of Use or Occupancy

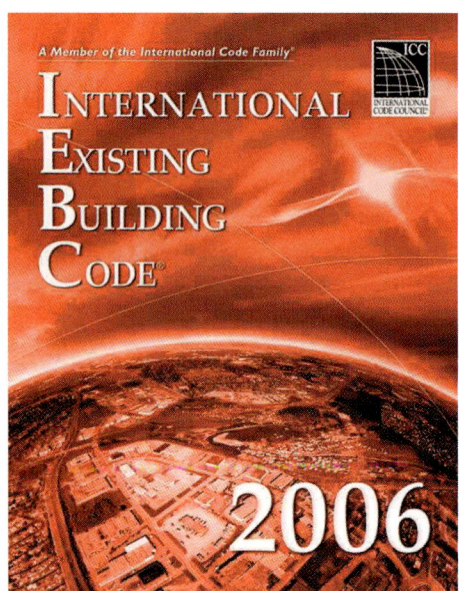

International Existing Building Code.

CHANGE TYPE. Modification

CHANGE SUMMARY. References to the International Existing Building Code (IEBC) for additions, alterations, and change of occupancy have been deleted from Sections 102.3 through 102.5. **(F8-04/05)**

2006 CODE: ~~**102.3 Change of Use or Occupancy.** The provisions of the *International Existing Building Code* shall apply to all buildings undergoing a change of occupancy.~~

~~**102.4. Application of Building Code.** The design and construction of new structures shall comply with the *International Building Code*. Repairs, alterations, and additions to existing structures shall comply with the *International Existing Building Code*.~~

~~**102.5 Historic Buildings.** The construction, alteration, repair, enlargement, restoration, relocation, or movement of existing building or structures that are designated as historic buildings when such buildings or structures do not constitute a distinct hazard to life or property shall be in accordance with the provisions of the *International Existing Building Code*.~~

102.3 Change of Use or Occupancy. No change shall be made in the use or occupancy of any structure that would place the structure in a different division of the same group or occupancy or in a different group of occupancies, unless such structure is made to comply with the requirements of this code and the *International Building Code*. Subject to the approval of the fire code official, the use or occupancy of an existing structure shall be allowed to be changed and the structure is allowed to be occupied for purposes in other groups without conforming to all the requirements of this code and the *International Building Code* for those groups, provided the new or proposed use is less hazardous, based on life and fire risk, than the existing use.

102.4 Application of Building Code. The design and construction of new structures shall comply with the *International Building Code*, and any alterations, additions, changes in use, or changes in structures required by this code, which are within the scope of the *International Building Code*, shall be made in accordance therewith.

102.5 Historic Buildings. The provisions of this code relating to the construction, alteration, repair, enlargement, restoration, relocation, or moving of buildings or structures shall not be mandatory for existing buildings or structures identified and classified by the state or local jurisdiction as historic buildings when such buildings or structures do not constitute a distinct hazard to life or property. Fire protection in designated historic buildings and structures shall be provided in accordance with an approved fire protection plan.

CHANGE SIGNIFICANCE. Provisions in these sections have been returned to the text found in the 2000 edition of the *International Fire*

Code (IFC). In the 2003 edition, the code had been revised to reference the IEBC as a basis of regulating changes made to existing buildings and occupancies.

Adoption of the 2003 IEBC was very controversial. Fire code officials participating in the IFC development process harbored concerns that the IEBC, as a new code, first published in 2003, was in need of further work to resolve perceived weaknesses in some of the provisions. For example, using the IEBC, some believed that it might be permissible to construct a shell building complying with the *International Building Code* (IBC), get a certificate of occupancy, and then finish out the occupancy based on the more lenient provisions in the IEBC because the finish-out work would be a modification of an "existing" building.

To address these concerns, references to the IEBC were removed from the 2006 IFC, which has the consequence of requiring additions, alterations, and changes of occupancy to comply with the requirements for new construction specified in the IBC, unless (1) the code official determines that the new use is less hazardous than the old use from a fire- and life-safety perspective, or (2) the structure is designated as a historic building. In the 2006 IFC, historic buildings are governed by applicable state/local regulations, provided that they do not constitute a distinct hazard to life or property and that an approved fire protection plan has been developed.

Table 105.6.20

Permit Amounts for Hazardous Materials

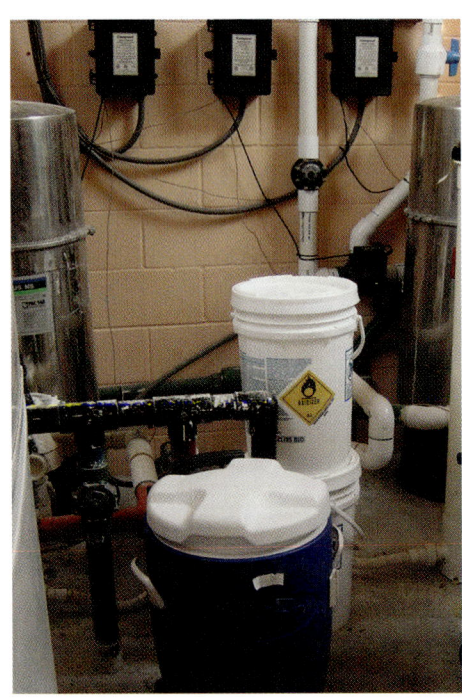

Class 3 oxidizers are commonly stored at apartment complexes for pool maintenance.

CHANGE TYPE.   Modification

CHANGE SUMMARY.   Threshold quantities for requiring a permit for Class 3 oxidizer solids and liquids have been increased. **(F6-03/04)**

2006 CODE:

TABLE 105.6.20   **Permit Amounts for Hazardous Materials**

Type of Material		Amount
Oxidizing materials		
Gases		See Section 105.6.8
Liquids		
	Class 4	Any Amount
	Class 3	1 gallon[a]
	Class 2	10 gallons
	Class 1	55 gallons
Solids		
	Class 4	Any Amount
	Class 3	10 pounds[b]
	Class 2	100 pounds
	Class 1	500 pounds

a. 20 gallons when Table 2703.1(1) Note k applies and hazardous material identification signs in accordance with Section 2703.5 are provided for quantities of 20 gallons or less.

b. 200 pounds when Table 2703.1(1) Note k applies and hazardous material identification signs in accordance with Section 2703.5 are provided for quantities of 200 pounds or less.

(Portions of the table not shown remain unchanged)

CHANGE SIGNIFICANCE.   In the past, threshold quantities triggering a permit for Class 3 oxidizers have been the same as the tabular values set forth in Table 2703.1.1(1), which establishes the maximum quantities of hazardous materials allowed in a single control area, commonly referred to as the MAQ.

Although the tabular values were in agreement in the 2003 code, Table 2703.1.1(1) Note "k" modified the MAQ for Class 3 oxidizers, increasing the basic MAQ from 10 pounds to 200 pounds when:

1. the materials are necessary for maintenance purposes, operation, or sanitation of equipment,

2. the storage containers and the manner of storage are approved by the code official, and

3. a firefighter warning placard in accordance with Section 2703.5, which references National Fire Protection Association (NFPA) Standard 704, is provided as required by Notes "a" and "b" to Table 105.6.20.

For this specific case, it was considered reasonable to allow the permit threshold quantity to be increased to match the actual MAQ.

Accordingly, this change fully correlates the permit threshold quantity with the MAQ for Class 3 oxidizers when the 200-pound MAQ applies, and in doing so, it eliminates the permit requirement for small apartments and similar occupancies that use limited quantities of Class 3 oxidizers for sanitizing swimming pools. Requiring a permit for such cases was considered excessive.

An example of a Class 3 oxidizer commonly affected by this provision is calcium hypochlorite, which is often used as a swimming pool sanitizer.

106.4

Missed Violations–Approvals

CHANGE TYPE.   Addition

CHANGE SUMMARY.   A section has been added to clarify that successful completion of an inspection does not constitute an approval of code violations that may have been missed by the inspector. **(F15-04/05)**

2006 CODE: **106.4 Approvals.** Approval as the result of an inspection shall not be construed to be an approval of a violation of the provisions of this code or of other ordinances of the jurisdiction. Inspections presuming to give authority to violate or cancel provisions of this code or of other ordinances of the jurisdiction shall not be valid.

CHANGE SIGNIFICANCE.   The section recognizes that it is impossible for a fire code inspector in the normal course of an inspection to identify every code violation that may exist. The intent of the change is to clarify that a code violation missed by an inspector does not become an acceptable condition simply because the violation was not identified.

In some cases, conditions violating the code may be repeatedly overlooked by inspectors, only to be picked up at some time in the future. Occupants may ask, "Why didn't anyone ever bring this up before?" or state, "It's been this way for years, and none of the other inspectors who have been here ever made an issue of this," expecting that they should be given an allowance to have a code violation continue since the condition has been in existence for some period of

Simply because a code violation may be missed by an inspector, that doesn't cause the condition to become legal.

time. This section gives an inspector a place in the code to point to, indicating that violations that have been missed in the past must still be corrected.

Similar provisions exist in IFC Section 105.3.6, which governs the issuance of permits and renders a permit invalid if it was erroneously issued for a condition that violated the code, and in IBC Section 109.1.

202

Definitions for Emergency Shutoff Controls and Valves

CHANGE TYPE.   Addition

CHANGE SUMMARY.   Several new definitions have been added to clarify requirements throughout the code regulating safety valves for liquid and gas handling systems. **(F8-03/04 and F19-04/05)**

2006 CODE: Remotely Located, Manually Activated Shutdown Control.   A control system that is designed to initiate shutdown of the flow of gases or liquids that is manually activated from a point located some distance from the delivery system.

Emergency Shutoff Valve.   A valve designed to shut off the flow of gases or liquids.

Emergency Shutoff Valve, Automatic.   A fail-safe automatic closing valve designed to shut off the flow of gases or liquids, initiated by a control system that is activated by automatic means.

Emergency Shutoff Valve, Manual.   A manually operated valve designed to shut off the flow of gases or liquids.

CHANGE SIGNIFICANCE.   In past editions, requirements for emergency shutoff valves have been inconsistently used in various sections of the code. The new definitions, along with many changes in other sec-

A Manual Emergency Shutoff Valve.

tions of the code that correlate with the defined terms, will help to clarify the types of valves that are intended for various service conditions.

There are fundamentally two types of emergency shutoff valves commonly required by code: manually operated and automatically operated. Manual valves require human intervention to operate. Such intervention may be initiated either remotely (Remotely Located Manually Activated Shutdown Control) or at the valve (Manual Emergency Shutoff Valve).

Automatic Emergency Shutoff Valves are automatically actuated by a control system. Although a typical automatic valve may fail in either the open or closed position, an Automatic Emergency Shutoff Valve is defined by code to require a design that fails in the "safe" position (see discussion of the new definition of FAIL-SAFE in Section 202 on page 11). Examples of failure conditions that would ordinarily be considered include a loss of electricity or pneumatic or hydraulic pressure to the valve controller. With a spring-operated fail-safe valve that fails in the closed position, the valve would be opened by electricity or pneumatic or hydraulic pressure, and the spring would force the valve closed upon loss of these.

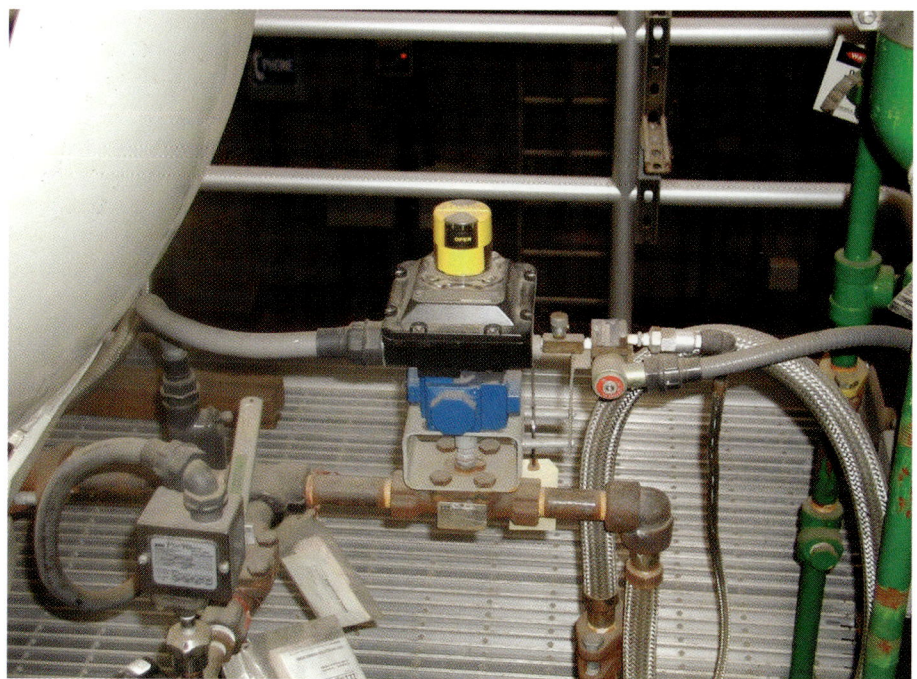

An Automatic Emergency Shutoff Valve.

202

Definition for Decorative Materials

CHANGE TYPE.   Addition

CHANGE SUMMARY.   A definition of DECORATIVE MATERIALS has been added to clarify application of the term, which is used in Chapters 3, 7, 8, and 24, but was previously undefined. **(F17-04/05)**

2006 CODE: Decorative Materials.   All materials applied over the building interior finish for decorative, acoustical, or other effect (such as curtains, draperies, fabrics, streamers, and surface coverings), all other materials utilized for decorative effect (such as: batting, cloth, cotton, hay, stalks, straw, vines, leaves, trees, moss, and similar items), including foam plastics and materials containing foam plastics; decorative materials do not include floor coverings, ordinary window shades, interior finish, and materials 0.025 inch (0.64 mm) or less in thickness applied directly to and adhering tightly to a substrate.

CHANGE SIGNIFICANCE.   Decorative materials are regulated by IFC Chapters 3, 7, 8, and 24, with primary regulations being located in Sections 806 (decorative vegetation) and 807 (decorative materials other than decorative vegetation). The new definition was based in part on the Standard Fire Prevention Code, with modifications having been made to correlate with the use of the term in the IFC.

Decorative materials that are not suitably fire-safe have been cited as being contributory to many major life-loss fires, particularly in assembly occupancies, where a rapidly growing fire can have catastrophic consequences. By adding a definition and rewriting Chapter 8 in the 2006 IFC, understanding and enforcement of regulations governing decorative materials should be improved.

Decorative materials, often appearing somewhat harmless to the untrained eye, can burn very rapidly if not in compliance with applicable code requirements. Decorative materials have been associated with many fires in which multiple lives were lost.

CHANGE TYPE.   Addition

CHANGE SUMMARY.   A definition of FAIL-SAFE has been added to clarify application of the term, which is used in Chapters 2, 27, 35, 37, and 40, but was previously undefined. **(F19-04/05)**

2006 CODE: Fail-Safe.   A design condition incorporating a feature for automatically counteracting the effect of an anticipated possible source of failure; also, a design condition eliminating or mitigating a hazardous condition by compensating automatically for a failure or malfunction.

CHANGE SIGNIFICANCE.   The term FAIL-SAFE is used in the definitions of EMERGENCY SHUTOFF VALVE, AUTOMATIC and EXCESS FLOW CONTROL in Chapter 2, as well as several locations in Chapters 27, 35, 37, and 40, which regulate equipment handling various hazardous materials. In most cases, the fail-safe condition would stop the movement or processing of hazardous materials, such as a valve that defaults to the closed position when a failure in the valve-control system occurs. However, some processes require staged shutdown to be safely stopped, and in these cases, a process hazard analysis is necessary to determine the fail-safe mode.

The new definition is based on a combination of a definition in Webster's Dictionary and one found in NFPA Standard 70E, which is the only NFPA document to define the term, even though it appears in many NFPA codes and standards.

202
Definition for Fail Safe

Example of a fail-safe valve that closes automatically in the event of a fault condition.

202

Definition for Assembly Group A

CHANGE TYPE.   Modification

CHANGE SUMMARY.   The definition of "Occupancy classification, Assembly Group A" has been revised to indicate that an assembly space having an occupant load of less than 50 persons is now permitted to be classified as a Group B occupancy when accessory to another occupancy. **(G24-04/05 and G33-04/05)**

2006 CODE: **[B] Assembly Group A.**   Assembly Group A occupancy includes, among others, the use of a building or structure, or a portion thereof, for the gathering together of persons for purposes such as civic, social or religious functions, recreation, food or drink consumption, or awaiting transportation. ~~A room or space used for assembly purposes by less than 50 persons and accessory to another occupancy shall be included as a part of that occupancy. Assembly areas with less than 750 square feet (69.7 m²) and which are accessory to another occupancy according to Section 302.2.1 of the *International Building Code* are not assembly occupancies. Assembly occupancies which are accessory to Group E in accordance with Section 302.2 of the *International Building Code* are not considered assembly occupancies. Religious educational rooms and religious auditoriums which are accessory to churches in accordance with Section 302.2 of the *International Building Code* and which have occupant loads of less than 100 shall be classified as A-3. A building or tenant space used for assembly purposes by less than 50 persons shall be considered a Group B occupancy.~~

Example:

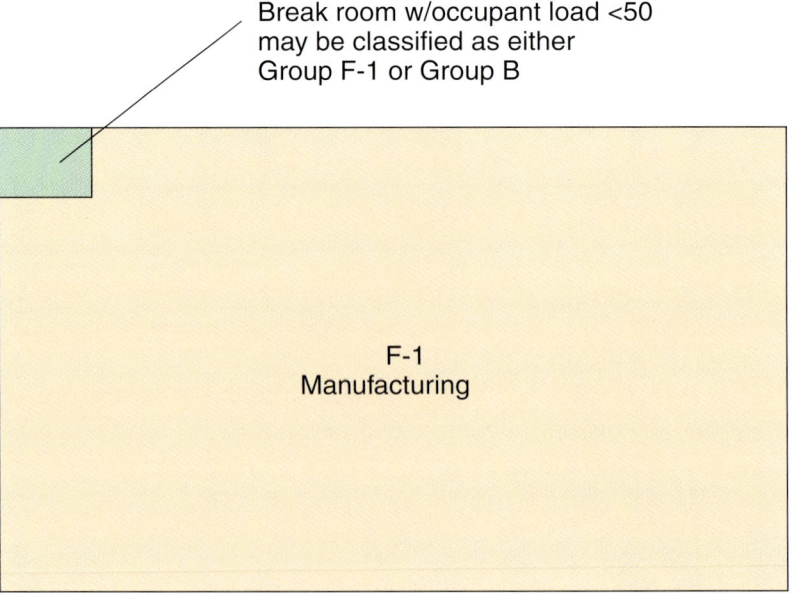

Break room w/occupant load <50 may be classified as either Group F-1 or Group B

F-1
Manufacturing

Small Assembly Uses

Exceptions:

1. A building used for assembly purposes with an occupant load of less than 50 persons shall be classified as a Group B occupancy.

2. A room or space used for assembly purposes with an occupant load of less than 50 persons, and accessory to another occupancy, shall be classified as a Group B occupancy or classified as part of that occupancy.

3. A room or space used for assembly purposes that is less than 750 square feet (70 m^2) in area and is accessory to another occupancy shall be classified as a Group B occupancy or classified as part of that occupancy.

(for remaining text, see the 2006 IFC)

CHANGE SIGNIFICANCE.   Where spaces used for assembly purposes have an occupant load not exceeding 49 persons or an area of less than 750 square feet (both considered small assembly areas), they are not to be classified as Group A occupancies. Rather, the code provides two different approaches to the proper classification of such uses:

1. If a small assembly area is not accessory to another occupancy in the building, or is freestanding, the appropriate classification is Group B.

2. Where the small assembly area is accessory to another occupancy in the building, one of two alternatives can be used. First, as was the case in earlier editions of the code, the assembly area can be classified with the same occupancy group as the use that it serves. The second alternative allows assigning a Group B classification to the assembly area, at the designer's option.

Provisions previously in the Group A definition permitting assembly occupancies associated with Group E to be considered as part of the Group E occupancy have been deleted. IBC Section 508.3.1, Exception 2, which states that no fire separation is required between Group E occupancies and associated assembly areas, makes those provisions no longer necessary.

Provisions previously in the Group A definition dealing with religious educational rooms and religious auditoriums have been deleted since these were relocated to the Group E definition in 2003. However, relocation of the religious auditorium provision was an apparent error since these uses should be classified as Group A-3 once they exceed an occupant load of 49.

It is also worth noting that IBC Section 508.3.1, Exception 3, permits a special allowance for no fire separation between religious educational rooms and religious auditoriums with an occupant load not exceeding 99 and areas to which they are accessory.

202

Definition for Business Group B

CHANGE TYPE.   Addition

CHANGE SUMMARY.   The definition of "Occupancy classification, Business Group B" has been revised by expanding the list of uses typically classified as Group B occupancies to include training and skill-development activities that are not associated with a school or academic program. **(G37-04/05)**

2006 CODE: **[B] Business Group B.**   Business Group B occupancy includes, among others, the use of a building or structure, or a portion thereof, for office, professional, or service-type transactions, including storage of records and accounts. Business occupancies shall include, but not be limited to, the following:

   <u>Training and skill development not within a school or academic program</u>

   (no changes to other listed items)

CHANGE SIGNIFICANCE.   Various types of facilities are designed to teach and train personnel outside of academic school programs, and the code has not previously been specific as to what occupancy classification should be assigned. These types of uses include various skill, trade, and technical programs. The addition of a specific listing for such uses as Group B occupancies addresses pertinent questions as to appropriate occupancy classification.

   The Group B classification of such training and skill programs is not limited to persons beyond the 12th grade. For example, a retail music instrument business often includes training for students who purchase instruments. On the basis of this code change, a business of this type is considered as a mercantile (Group M) occupancy with an accessory business (Group B) occupancy where classes are conducted. Similar uses, such as tutoring businesses, are also included in the Group B classification.

Music schools associated with retail occupancies are examples of training and skill-development uses that are classified as Group B.

202

Definition for High-Hazard Group H, Exceptions

CHANGE TYPE. Modification

CHANGE SUMMARY. The definition of HIGH-HAZARD GROUP H has been revised by specifying occupancies and conditions that qualify for special exemptions from Group H and by better correlating with the requirements for control areas in Chapter 27. **(F193-03/04, F248-04/05)**

2006 CODE: High-Hazard Group H. High-hazard Group H occupancy includes, among others, the use of a building or structure, or a portion thereof, that involves the manufacturing, processing, generation or storage of materials that constitute a physical or health hazard in quantities in excess of ~~those found in Tables 307.7(1) and 307.7(2) of the *International Building Code*. (See also definition of "Control area")~~ quantities allowed in control areas constructed and located as required in Section 2703.8.3. Hazardous uses are classified in Groups H-1, H-2, H-3, H-4, and H-5 and shall be in accordance with this code and the requirements of Section 415 of the *International Building Code*.

> **Exceptions:** ~~Occupancies as provided for in the *International Building Code*~~ The following shall not be classified in Group H, but shall be classified in the occupancy that they most nearly resemble:
>
> 1. Buildings and structures that contain not more than the maximum allowable quantities per control area of hazardous materials as shown in Tables 2703.1.1(1) and 2703.1.1(2), provided that such buildings are maintained in accordance with this code.
> 2. Buildings utilizing control areas in accordance with Section 2703.8.3 that contain not more than the maximum allowable quantities per control area of hazardous materials as shown in Tables 2703.1.1(1) and 2703.1.1(2).

202 continues

Hazardous materials in dry cleaning equipment sometimes qualify for a special exemption from Group H occupancy requirements.

Hazardous materials in refrigeration systems qualify for a special exemption from Group H occupancy requirements.

202 continued

3. Buildings and structures occupied for the application of flammable finishes, provided that such buildings or areas conform to the requirements of Section 416 of the *International Building Code* and Chapter 15 of this code.

4. Wholesale and retail sales and storage of flammable and combustible liquids in mercantile occupancies conforming to Chapter 34.

5. Closed piping systems containing flammable or combustible liquids or gases utilized for the operation of machinery or equipment.

6. Cleaning establishments that utilize combustible liquid solvents having a flash point of 140° F (60° C) or higher in closed systems employing equipment listed by an approved testing agency, provided that this occupancy is separated from all other areas of the building by 1-hour fire barriers constructed in accordance with Section 706 of the *International Building Code* or 1-hour horizontal assemblies constructed in accordance with Section 711 of the *International Building Code,* or both.

7. Cleaning establishments that utilize a liquid solvent having a flash point at or above 200°F (93°C).

8. Liquor stores and distributors without bulk storage.

9. Refrigeration systems.

10. The storage or utilization of materials for agricultural purposes on the premises.

11. Stationary batteries utilized for facility emergency power, uninterrupted power supply or telecommunication facilities, provided that the batteries are provided with safety venting caps and ventilation is provided in accordance with the *International Mechanical Code.*

12. Corrosives shall not include personal or household products in their original packaging used in retail display or commonly used building materials.

Aerosol storage areas are among a number of special uses that qualify for special exemption from classification as a Group H occupancy.

13. Buildings and structures occupied for aerosol storage shall be classified as Group S-1, provided that such buildings conform to the requirements of Chapter 28.

14. Display and storage of nonflammable solid and nonflammable or noncombustible liquid hazardous materials in quantities not exceeding the maximum allowable quantity per control area in Group M or S occupancies complying with Section 2703.8.3.5.

15. The storage of black powder, smokeless propellant and small arms primers in Groups M and R-3 and special industrial explosive devices in Groups B, F, M, and S, provided such storage conforms to the quantity limits and requirements of this code.

(for remaining text, see the 2006 IFC)

CHANGE SIGNIFICANCE. Although the revised IFC definition for HIGH-HAZARD GROUP H appears to have been significantly changed, the changes are largely editorial, providing better coordination with IBC Sections 307 and 415 and IFC Chapter 27. The 15 exceptions listed with the definition of Group H are not new. They previously appeared in IBC Section 307.9, but they were often overlooked because they appeared at the end of Section 307 rather than at the beginning. They were not published at all in previous editions of the IFC.

The exceptions still appear in the 2006 IBC, but they were relocated to Section 307.1. The conditions listed in the exceptions are exempted from Group H on the basis of unique controls or conditions that were judged to satisfactorily address associated hazards.

An important change to the exceptions that took place between the 2003 provisions (published in IBC Section 307.9) and the 2006 provisions (published in IBC Section 307.1 and IFC Section 202) involves Exception 5. The change closes a significant loophole in the Group H occupancy requirements in previous editions. This exception previously stated "5. Closed systems housing flammable or combustible liquids or gases utilized for the operation of machinery or equipment." Now it states "5. Closed piping systems containing flammable or combustible liquids or gases utilized for the operation of machinery and equipment." The difference is the addition of the words "piping" and "containing" and the deletion of the word "housing."

Although not the intent of the code, Exception 5, as previously written could be interpreted to allow unlimited quantities of flammable or combustible liquids or gases on any floor in any occupancy outside of a High Hazard Group H-3 occupancy, provided that the system was a closed system utilized for the operation of equipment or machinery, such as a standby generator. As revised, the code is now clear that flammable and combustible liquids and flammable gases contained within closed piping systems are not counted when determining the quantities of hazardous materials in use. However, quantities of hazardous materials in storage tanks, such as generator day tanks for example, and other portions of systems must be counted.

In addition to revising Exception 5 to clarify the intent of the code with respect to closed systems, Table 2703.1.1(1) has been revised in the 2006 edition with the addition of a new footnote "p," which addresses other special hazardous materials quantity exclusions. Additional information on that revision is provided in the discussion of changes to Table 2703.1.1(1) herein.

202

Definition for High-Hazard Group H-3

Class 3 oxidizers in low-pressure storage and closed-use conditions are associated with H-3 occupancies, rather than H-2, when maximum allowable quantities per control area are exceeded.

CHANGE TYPE.   Clarification

CHANGE SUMMARY.   The Group H-3 occupancy classification category has been revised to include flammable and combustible liquids used or stored in normally closed containers or systems at exactly 15 psi. In addition, storage and closed use of oxidizers at 15 psi or less is now associated with the Group H-3 occupancy classification. The change related to oxidizers actually occurred in the 2003 edition but was not published in the first printings of that edition; however, it was later included in the 2003 edition errata. **(F20-04/05)**

2006 CODE: High-Hazard Group H-3.   Buildings and structures that contain materials that readily support combustion or present a physical hazard shall be classified as Group H-3. Such materials shall include, but not be limited to, the following:

Class I, II or IIIA flammable or combustible liquids which are used or stored in normally closed containers or systems pressurized at ~~less than~~ 15 psi (103 kPa) gauge <u>or less</u>.

Combustible fibers

Consumer fireworks, 1.4G (Class C Common)

Cryogenic fluids, oxidizing

Flammable solids

Organic peroxides, Classes II and III

<u>Oxidizers, Class 3, that are used or stored in normally closed containers or systems pressurized at 15 pounds per square inch gauge (103 kPa) or less.</u>

Application of the 15 psi breakpoint between Group H-2 and Group H-3 for pressurized containers and systems containing flammable and combustible liquids has been clarified.

Oxidizers, Class 2

Oxidizing gases

Unstable (reactive) materials, Class 2

Water-reactive materials, Class 2

CHANGE SIGNIFICANCE.   In the 2003 edition, the code was changed to place oxidizers and flammable liquids at an equivalent level with respect to occupancy classifications. All open-use conditions and those storage and closed-use conditions involving vessels or systems pressurized in excess of 15 psig were linked to the H-2 occupancy classification. Storage and closed-use conditions in vessels or systems having a pressure less than 15 psig were linked to the H-3 occupancy classification. However, neither the H-2 nor the H-3 provisions addressed the case of vessels or systems with exactly 15 psig, presenting an anomaly in the code. This was addressed by modifying the H-3 definition to include the 15 psig case.

It should be noted that the provisions for storage and closed use of Class 3 oxidizers at pressures not exceeding 15 psig were approved for publication in the 2003 IFC but were inadvertently omitted from early printings of that code. They were later added by errata and now appear in the 2006 edition. The text associated with this revision is denoted in the "2006 Code" section by text shown with a dashed underline.

The original basis for the 15 psig threshold is American National Standards Institute (ANSI) Standard B31.3, Section 300.1.3, which excludes piping that operates below that pressure and carrying certain nonhazardous materials from having to comply with ANSI B31.3. Because ANSI B31.3 uses 15 psig as a suitable breakpoint for application of that standard, that pressure was selected as a basis for applicability of selected IFC hazardous materials regulations, including the breakpoint for H-2 versus H-3 occupancy classification for uses containing sufficient quantities of flammable and combustible liquids or oxidizers.

202

Definition for Residential Group R

CHANGE TYPE.   Addition

CHANGE SUMMARY.   The definition of "Occupancy classification, Residential Group R" has been revised to correlate with a new definition of "congregate living facilities" in the IBC. The occupancy classification of such uses is now clearly dependent on the number of persons housed. Where the facility has 16 or fewer occupants, it is considered a Group R-3 occupancy; otherwise, the previous classification of Group R-2 remains appropriate. **(G42–04/05)**

2006 CODE: [B] Residential Group R.   Residential Group R includes, among others, the use of a building or structure, or a portion thereof, for sleeping purposes when not classified as an Institutional Group I <u>or when not regulated by the *International Residential Code* in accordance with Section 101.2 of the *International Building Code.*</u> Residential occupancies shall include the following:

R-1 (no change to text)

R-2 Residential occupancies containing sleeping units or more than two dwelling units where the occupants are primarily permanent in nature, including:

> Apartment houses
>
> Boarding houses (not transient)
>
> Convents
>
> Dormitories
>
> Fraternities and sororities
>
> Hotels (nontransient)
>
> Monasteries

Occupancy classification requirements for congregate living facilities, such as group homes, have been improved.

Motels (nontransient)

Vacation timeshare properties

Congregate living facilities with 16 or fewer occupants are permitted to comply with the construction requirements for Group R-3.

R-3 Residential occupancies where the occupants are primarily permanent in nature and not classified as Group R-1, R-2, R-4, or I, ~~and where~~ including:

Buildings that do not contain more than two dwelling units. ~~as applicable in Section 101.2, or~~

Adult care facilities that provide accommodations for five or fewer persons of any age for less than 24 hours. ~~and~~

Child care facilities that provide accommodations for five or fewer persons of any age for less than 24 hours.

Congregate living facilities with 16 or fewer persons.

Adult and child care facilities that are within a single-family home are permitted to comply with the *International Residential Code.* ~~in accordance with Section 101.2.~~

R-4 (no change to text)

CHANGE SIGNIFICANCE. In the classification of buildings containing sleeping units where the occupants are primarily permanent in nature, Group R-2 occupancies have traditionally included various types of congregate living facilities. Such uses include fraternity and sorority houses, dormitories, convents, monasteries, and boarding houses. A new definition of CONGREGATE LIVING FACILITIES has been added in the IBC, which defines these facilities to include buildings or parts thereof that contain sleeping units where residents share bathroom and/or kitchen facilities. This new definition helps to distinguish congregate living facilities versus other Group R-2 and R-3 occupancies.

In the 2006 edition, some congregate living facilities continue to be classified as Group R-2 occupancies, but facilities having an occupant load of 16 or fewer are now classified as Group R-3 occupancies. Small boarding houses, convents, monasteries, dormitories, fraternity houses, sorority houses, and nontransient hotels and motels are examples of congregate living facilities that may fall into the Group R-3 classification.

It was deemed that facilities with a limited occupant load operate in a manner very similar to single-family dwelling and should be classified accordingly. The threshold of 16 occupants is based on the criteria for Group R-4 occupancies. Because the Group R-4 classification provides for a level of safety consistent with that for Group R-3 occupancies, the limit of 16 persons is appropriate. Larger occupant loads result in more hazardous conditions that are better addressed by the Group R-2 requirements.

202

Definition for Moderate-Hazard Storage, Group S-1

CHANGE TYPE.   Addition

CHANGE SUMMARY.   The definition of "Occupancy Classification, Moderate-hazard Storage, Group S-1" has been revised by expanding the list of uses typically classified as Group S-1 occupancies to include indoor dry boat storage. **(G46-03/04)**

2006 CODE: **[B] Moderate-Hazard Storage, Group S-1.**   Buildings occupied for storage uses which are not classified as Group S-2, including, but not limited to, storage of the following:

Dry boat storage (indoor)
(no changes to other listed items)

CHANGE SIGNIFICANCE.   This change specifically identifies dry boat storage within a building as a Group S-1 occupancy. The contents have some degree of combustibility, and as such they cannot be considered as Group S-2. In addition, the specific listing clarifies that dry boat storage should not be equated with a Group S-2 open or enclosed parking garage. Where the boat storage occurs outside of a building, there is no occupancy classification.

Group S-1 occupancies now include indoor dry boat storage.

PART **2**

General Safety Requirements

Chapters 3 and 4

- **Chapter 3**   General Precautions against Fire
- **Chapter 4**   Emergency Planning and Preparedness

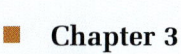 basic requirement of the *International Fire Code* is to prevent the unintended ignition of materials inside and outside of buildings. Controlling fuels and ignition sources are two methods that limit the potential for fire. Chapter 3 contains requirements for combustible waste materials, control or elimination of ignition sources, open flames and recreational fires and the use of smoking materials.

Certain equipment can also be a source of ignition, and Chapter 3 also addresses diverse topics such as the proper operation of asphalt kettles and powered industrial trucks. In occupancies such as assembly uses and covered malls, unique controls are specified for certain hazardous materials and displays of vehicles. Another general safety requirement in Chapter 3 is the construction specification for the protection of equipment from vehicular impact.

Chapter 4 addresses emergency planning and preparedness, covering such topics as fire safety plans and hazardous materials communications programs. ■

308.3.8

Open Flame Ignition Sources in Group R-2 Dormitories

309.5

Refueling of Powered Industrial Trucks and Equipment

311.5

Placards for Unsafe Buildings

313.1

Storage of Fueled Equipment in Buildings

404.2

Fire Safety and Evacuation Plans for Group B

404.2

Fire Safety and Evacuation Plans for Group R-2

TABLE 405.2

Frequency of Emergency Evacuation Drills

408.3

Emergency Planning and Preparedness for Group E Occupancies and Group R-2 College and University Buildings

308.3.8

Open Flame Ignition Sources in Group R-2 Dormitories

Candles and other common sources of fire ignition are prohibited in dormitories.

CHANGE TYPE.   Addition

CHANGE SUMMARY.   This new section introduces a prohibition of candles and similar ignition sources commonly used in dormitories. (F13-03/04)

2006 CODE: 308.3.8 Group R-2 Dormitories. Candles, incense, and similar open-flame–producing items shall not be allowed in sleeping units in Group R-2 dormitory occupancies.

CHANGE SIGNIFICANCE.   Fire represents a significant risk to life and property in dormitory occupancies, particularly at schools, colleges, and universities. The large number of individuals living in close proximity to one another creates the potential for a relatively small fire to have serious and fatal consequences. This new section gives code officials the authority to prohibit candles, incense, and similar items in these occupancies.

NFPA data cited to support this revision showed an average of 1425 fires each year in dormitories, causing $6.3 million in direct property damage. The data also indicated that candle-initiated fires cause losses of $2.3 million annually in direct property damage in these occupancies, which makes candle fires the most costly type of fire in a residence hall.

By prohibiting the use of candles, incense, and similar open-flame items, the goal is to reduce these losses. It is recognized that enforcement of this provision by fire inspectors will be challenging, but at a minimum, the regulation can serve as a useful basis for a dormitory operator to regulate occupant behavior and discipline violators.

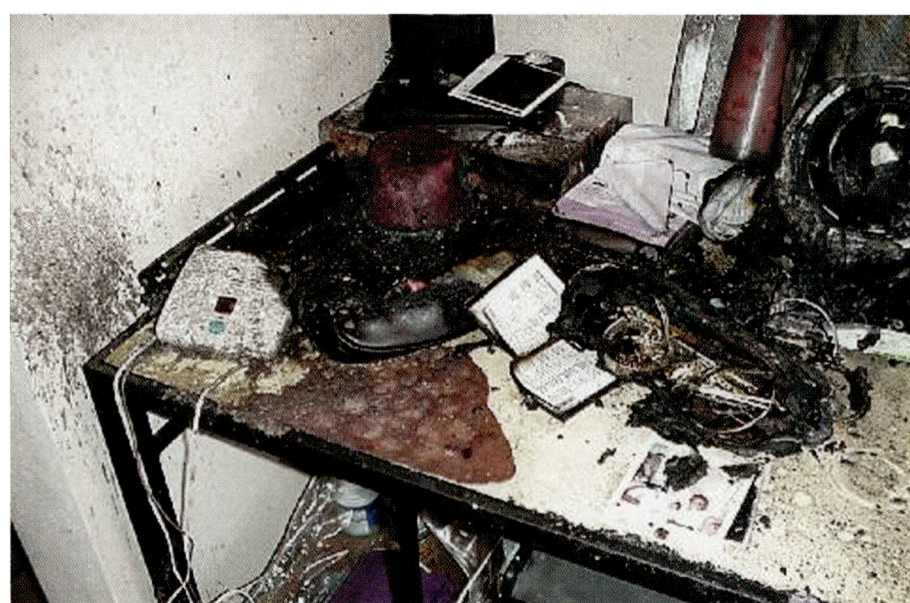

A candle started this desktop fire, which was fortunately extinguished before causing major damage.

CHANGE TYPE. Modification

CHANGE SUMMARY. The revised provisions clarify that fueling of forklifts and similar equipment using fixed motor fuel–dispensing equipment constitutes a motor fuel–dispensing operation, governed by Chapter 22. **(F14-03/04)**

2006 CODE: 309.5 Refueling. Powered industrial trucks using liquid fuel or LP-gas shall be refueled outside of buildings or in areas specifically approved for that purpose. Fixed fuel-dispensing equipment and associated fueling operations shall be ~~and~~ in accordance with Chapter 22. Other fuel-dispensing equipment and operations, including cylinder exchange for LP-gas fueled vehicles, shall be in accordance with Chapter 34 for flammable and combustible liquids or Chapter 38 for LP-gas.

CHANGE SIGNIFICANCE. This section in the 2003 IFC implied that filling of forklifts and other powered industrial trucks was not "dispensing of fuel into the tank of a motor vehicle," which is the trigger for an operation to be regarded as a motor fuel–dispensing operation governed by Chapter 22 when fixed equipment is involved. The revised text makes the application of this section clear, requiring compliance with Chapter 22 when fixed equipment is involved. Compliance with Chapter 34 is required for use of portable fueling equipment for flammable and combustible liquid fuels and Chapter 38 for exchange of fuel cylinders containing liquefied petroleum (LP) gas.

309.5

Refueling of Powered Industrial Trucks and Equipment

When forklifts and similar equipment are fueled with fixed equipment, the fixed equipment must comply with code requirements for motor fuel–dispensing stations in Chapter 22.

311.5

Placards for Unsafe Buildings

CHANGE TYPE. Addition

CHANGE SUMMARY. A new section has been added to standardize requirements for placarding of buildings that have been determined to be unsafe in accordance with IFC Section 110. **(F28-04/05)**

2006 CODE: 311.5 Placards. Any building or structure determined to be unsafe pursuant to Section 110 of this code shall be marked as required by Sections 311.5.1 through 311.5.5.

311.5.1 Placard Location. Placards shall be applied on the front of the structure and be visible from the street. Additional placards shall be applied to the side of each entrance to the structure and on penthouses.

311.5.2 Placard Size and Color. Placards shall be 24 inches by 24 inches in size with a red background and white reflective stripes and a white reflective border. The stripes and border shall have a two inch stroke.

311.5.3 Placard Date. Placards shall bear the date of their application to the building and the date of the most recent inspection.

311.5.4 Placard Symbols. The design of the placards shall use the following symbols:

1. ☐ This symbol shall mean that the structure had normal structural conditions at the time of marking.

New regulations standardize placards for buildings that are determined to be unsafe. (Photo courtesy of Chris McKal, New Orleans, Louisiana, Fire Department Photo Unit.)

2. ⊡ This symbol shall mean that structural or interior hazards exist and interior firefighting or rescue operations should be conducted with extreme caution.

3. ⊠ This symbol shall mean that structural or interior hazards exist to a degree that consideration should be given to limit firefighting to exterior operations only, with entry only occurring for known life hazards.

311.5.5 Informational Use. The use of these symbols shall be informational only and shall not in any way limit the discretion of the on-scene incident commander.

CHANGE SIGNIFICANCE. This revision is intended to create a nationally recognized marking system to identify the degree of structural stability of vacant buildings for firefighters who may be called upon to fight a fire therein. The new requirement is in response to a fire in Worcester, Massachusetts, where six firefighters were killed in the line of duty while fighting a fire in an abandoned cold-storage warehouse in 1999. A recommendation for the placarding system was contained in the subsequent fire investigation report entitled *Abandoned Cold Storage Warehouse Multi-Firefighter Fatality Fire,* which was published by the Federal Emergency Management Agency.

Even with these provisions in the IFC, creation of a uniform marking system for nationwide use will prove difficult because many jurisdictions have already adopted their own local or statewide placarding programs for buildings that are abandoned or that have been damaged in a natural disaster. Nevertheless, the IFC provides a national model for jurisdictions choosing to adopt this section of the code.

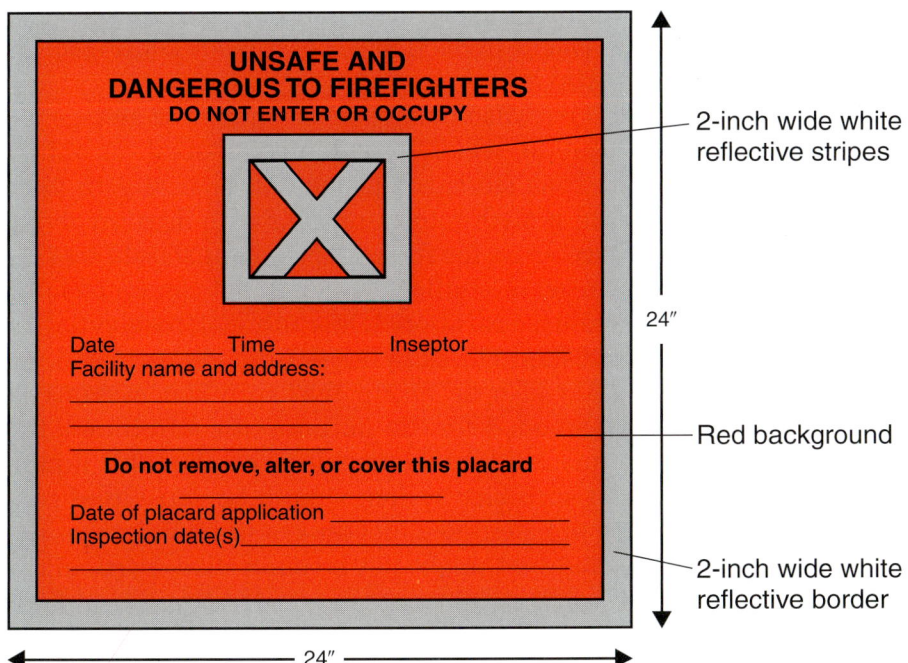

313.1

Storage of Fueled Equipment in Buildings

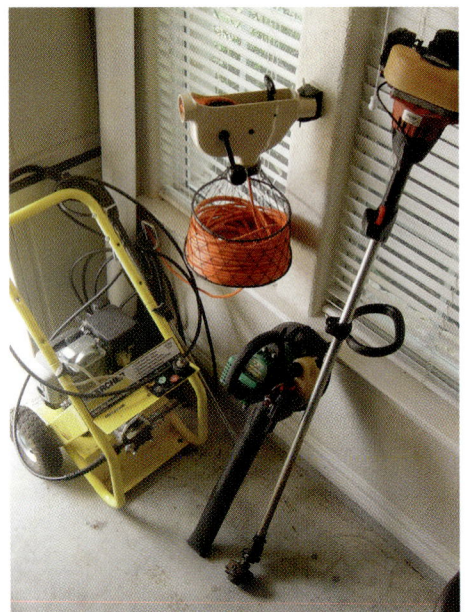

Small quantities of fuel for maintenance purposes are now allowed in sprinklered buildings.

CHANGE TYPE.   Modification

CHANGE SUMMARY.   A new exception has been added to permit small quantities of fuel for maintenance purposes to be located in sprinklered buildings. **(F29-04/05)**

2006 CODE: 313.1 Fueled Equipment.   Fueled equipment, including but not limited to motorcycles, mopeds, lawn-care equipment and portable cooking equipment, shall not be stored, operated or repaired within a building.

> **Exceptions:**
> 1. Buildings or rooms constructed for such use in accordance with the *International Building Code.*
> 2. Where allowed by Section 314.
> 3. Storage of equipment utilized for maintenance purposes is allowed in approved locations when the aggregate fuel capacity of the stored equipment does not exceed 10 gallons and the building is equipped throughout with an automatic sprinkler system installed in accordance with Section 903.3.1.1.

CHANGE SIGNIFICANCE.   The new Exception 3 provides a limited allowance for storing fueled equipment in sprinklered buildings. The change was justified on the basis of parity with Section 3404.3.4.4, which permits storage of up to 10 gallons of flammable or combustible liquid for maintenance purposes or operation of equipment in any occupancy without requiring a liquid storage cabinet.

Although Section 3404.3.4.4 is not intended to regulate fuel in fuel tanks, which instead falls under Section 313, it was considered to constitute a reasonable basis for relaxing the provisions of Section 313. Recognizing that fuel in equipment fuel tanks is at greater risk of spillage than flammable liquids kept in closed containers because of equipment wear, the allowance in Section 313 was made contingent on the storage area being sprinklered.

The new Exception 3 was also made contingent on the storage being in an "approved" location, which gives a fire inspector the ability to limit locations where such storage is permitted. Logically, in view of the intensity and rapid growth rate associated with fires involving flammable liquids, storage areas for fueled equipment should be limited to garages, utility areas, or similar locations that can be isolated from areas used for assembly, educational, institutional, residential, or other purposes with a substantial risk to life or property.

Fire inspectors should also be aware that sprinkler systems in most buildings are not intended for the purpose of controlling a fire fueled with significant quantities of flammable or combustible liquid. Thereby, while a typical sprinkler system should be capable of controlling an exposure fire threatening a fuel tank, a fire involving a significant quantity of spilled fuel might not be controlled, particularly in an area where residential sprinklers are provided because of their limited discharge rate and unique discharge pattern as compared to standard-spray commercial sprinklers.

It should also be noted that application of Exception 3 to equipment fueled with flammable gas, such as an LP gas–fueled cooking appliance, is beyond the intended scope of the exception, even though one might read the exception to imply that 10 gallons of liquefied gas fuel might be permitted. Because the exception was justified on the basis of parity with Chapter 34, it is clear that the intended application of the exception was limited to flammable and combustible liquids.

404.2

Fire Safety and Evacuation Plans for Group B

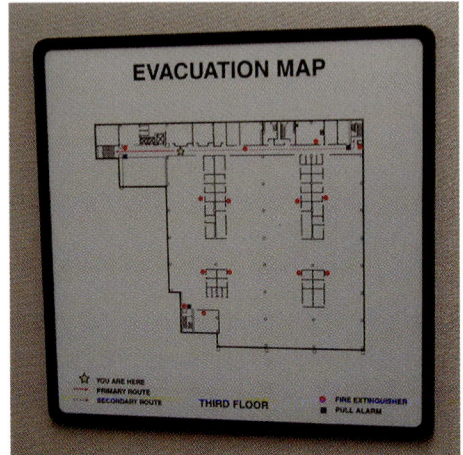

Group B buildings with an occupant load of 500 or more, or more than 100 persons above or below the lowest level of exit discharge, are now required to prepare and maintain an approved fire safety and evacuation plan.

CHANGE TYPE.   Addition

CHANGE SUMMARY.   A requirement has been added to mandate fire and evacuation plans for large Group B occupancies. A correlating change has been made to Table 405.2 to require fire and evacuation drills. **(F34-04/05)**

2006 CODE: 404.2 Where Required.   An approved fire safety and evacuation plan shall be prepared and maintained for the following occupancies and buildings.

1. Group A, other than Group A occupancies used exclusively for purposes of religious worship that have an occupant load less than 2000.
2. <u>Group B buildings having an occupant load of 500 or more persons or more than 100 persons above or below the lowest level of exit discharge.</u>

(no other changes to listed items)

CHANGE SIGNIFICANCE.   This change correlates fire alarm requirements in Section 907.2.2 with fire safety and evacuation plan requirements in this section by establishing similar triggers for both provisions with respect to Group B occupancies. Section 907.2.2 requires a fire alarm system when the occupant load of a Group B occupancy is either 500 or more in total or 100 or more above or below the lowest level of exit discharge. Alarm initiation is required to be by either sprinkler water flow in sprinklered buildings or manual pull stations in nonsprinklered buildings.

The philosophical basis for the change was a belief that once a Group B occupancy requires the installation of fire alarm notification appliances, fire safety and evacuation plans should prescribe an appropriate response.

There are a couple of possible points of confusion with respect to applying this section. First, the section refers to "Group B buildings," as opposed to "Group B occupancies," which implies that the provision applies only if the entire building is used for Group B. This would not be a proper application of the code. The intent of the proponent of this change was to correlate Sections 907.2.2 with Section 404.2, and because Section 907.2.2 uses "Group B occupancies" as the basis for triggering fire alarm requirements, Section 404.2 should be read in the same way, i.e., "Group B occupancies" as opposed to "Group B buildings." The proponent of this change simply chose the wrong terminology when the proposal was written, and this was not corrected when the change was processed.

A second possible point of confusion relates to the phrase "above or below the lowest level of exit discharge," which is used in both Section 907.2.2 and Section 404.2. This phrase is defined in Section 1002 to mean "the horizontal plane located at the point at which an exit terminates and an exit discharge begins." Technically, based on this definition, the lowest level above the exit discharge for a building constructed as slab on grade is the first story because the lowest level of exit discharge is the horizontal plane of the floor and the first story is above that plane. To interpret Sections 907.2.2 and 404.2 in this manner would be inconsistent with the intent of these provisions, which was to look at the "lowest level of exit discharge" as being the lowest story where an exit discharge is located. Proper application of these sections would be to apply the 100-occupant threshold to stories above or below the lowest story of exit discharge.

In addition to the new requirement in Section 404.2 to develop and maintain fire and emergency evacuation plans for some Group B occupancies, a related provision was added to Table 405.2 in the 2006 edition to require that these plans be exercised in drills on an annual basis. Additional information on the change to Table 405.2 is included in the discussion of that section herein.

404.2

Fire Safety and Evacuation Plans for Group R-2

CHANGE TYPE.   Addition

CHANGE SUMMARY.   A requirement has been added to mandate fire and evacuation plans for Group R-2 college and university buildings. Correlating changes have been made to Table 405.2 and Section 408.3 to require fire and evacuation drills. **(F35-04/05)**

2006 CODE: 404.2 Where Required.   An approved fire safety and evacuation plan shall be prepared and maintained for the following occupancies and buildings.

(No changes to Items 1–6)

<u>7.</u> Group R-2 college and university buildings.

(No changes to Items 8–13)

CHANGE SIGNIFICANCE.   In response to several major fires that have occurred in college dormitories and in fraternity and sorority housing, a requirement has been added to mandate fire safety and evacuation plans for Group R-2 college and university buildings. The proposal that resulted in this change also originally recommended requiring fire and evacuation plans for all Group B college and university buildings, but that portion of the proposal was rejected because the justification offered to support the change was considered to be inadequate. Note however that the previous change to Section 404.2 picked up this requirement for large Group B uses.

Because of the life-safety risks and unusual fuel loads that may be found in dormitory rooms, the IFC requires R-2 college and university buildings to prepare and maintain approved fire safety and evacuation plans and to conduct fire drills.

The goal of this new requirement is to help ensure that occupants in college and university dormitories are familiar with proper emergency egress procedures. College and university dormitories are often large buildings, sometimes high rise, that present unique challenges in ensuring timely evacuation due to the relatively young age of occupants and the types of activities that may take place in these occupancies.

A possible point of confusion with respect to applying this section involves the use of the terminology "Group R-2 college and university buildings," which implies that the provision applies only if the entire building is used for Group R-2. Interpreting the text in this manner would not result in proper application of the code. To be consistent with the intent of the code, the text should be read as "Group R-2 occupancies in college and university buildings." The proponent of this change simply chose the wrong terminology when the proposal was written, and this was not corrected when the change was processed.

Another possible point of confusion involves the reference to "college and university buildings." Although on-campus housing is clearly encompassed by the code, off-campus housing is less clearly addressed. It could be argued, and would be consistent with the code's intent, that off-campus fraternity and sorority housing is governed by this section. However, privately owned apartments that rent space to college and university students fall into a gray area. Such occupancies may have a few or many occupants who are not affiliated with a college or university. Thereby, these uses would be difficult to differentiate from any other R-2 occupancy. Because the code is somewhat vague with respect to these situations, enforcement becomes discretionary, and it is recommended that jurisdictions adopt a policy on application of this section to off-campus housing. See IFC Section 104.1 for guidance on establishing local policies.

404.2 continues

Rooms in college dormitories are often found to have unusual fire loads.

404.2 continued

For additional information on college and university dormitory safety, including a log of significant fires in these occupancies, see the Web site for the Center for Campus Fire Safety at www.campusfire.org.

In addition to the new requirement in Section 404.2 to develop and maintain fire and emergency evacuation plans for Group R-2 college and university buildings, related provisions have been added to Table 405.2 and Section 408.3 in the 2006 edition. Provisions in Table 405.2 require that fire and emergency evacuation plans be exercised in drills at least four times per year. Provisions in Section 408.3 provide additional details on conducting such drills. Additional information on the changes to Table 405.2 and Section 408.3 are included in the discussions of those sections herein.

Table 405.2

Frequency of Emergency Evacuation Drills

CHANGE TYPE.   Addition

CHANGE SUMMARY.   Requirements have been added to mandate fire and evacuation drills for large Group B occupancies, Group R-2 college and university buildings, and high-rise buildings. These provisions correlate with changes made to sections 404.2 and 408.3 in the 2006 edition. **(F34-04/05, F35-04/05, F37-04/05)**

2006 CODE:

TABLE 405.2   Fire and Evacuation Drill Frequency and Participation

Group or Occupancy	Frequency	Participation
Group A	Quarterly	Employees
Group B[c]	Annually	Employees
Group E	Monthly[a]	All occupants
Group I	Quarterly on each shift	Employees[b]
Group R-1	Quarterly on each shift	Employees
Group R-2[d]	Four annually	All occupants
Group R-4	Quarterly on each shift	Employees[b]
High-rise buildings	Annually	Employees

a. The frequency shall be allowed to be modified in accordance with Section 408.3.2.

b. Fire and evacuation drills in residential care assisted living facilities shall include complete evacuation of the premises in accordance with Section 408.10.5. Where occupants receive habilitation or rehabilitation training, fire prevention and fire safety practices shall be included as part of the training program.

c. Group B buildings having an occupant load of 500 or more persons or more than 100 persons above or below the lowest level of exit discharge.

d. Applicable to Group R-2 college and university buildings in accordance with Section 408.3.

Fires in high-rise buildings present unique dangers to occupants, making fire and evacuation drills essential components of a comprehensive fire-safety plan. Photo courtesy of Orange County, CA, Fire Authority.

CHANGE SIGNIFICANCE.   Three requirements mandating fire and emergency evacuation drills have been added. These new provisions, which impact large Group B occupancies, Group R-2 college and university buildings, and high-rise buildings, correlate with changes made to Sections 404.2 (Items 2 and 7) and 408.3. Fire and emergency evacuation drill frequencies specified in Table 405.2 are applicable only when fire and evacuation plans are required by Section 404.2 or when they are required by the code official in accordance with Section 405.1. More frequent drills can also be required by the code official in accordance with Section 405.2.

Group B Occupancies.   With respect to Group B occupancies, only occupancies that are within the scope of Table 405.2, Note c, with occupant-load thresholds of 100 and 500 are covered by Table 405.2. For guidance on how to interpret these thresholds, see the discussion on Group B occupancies under Section 404.2 herein. Group B occupancies required to conduct fire and evacuation drills must do so at least annually, and all employees must participate. The term "employees" in

Table 405.2 continues

Table 405.2 continued

the Group B requirement is intended to encompass anyone who works in the occupancy on a regular basis, even if their job function has nothing to do with management or operation of the occupancy, so tenants would be included. This differs from use of the term "employees" in the table entries for Groups A, I, R-1, and R-4. In these occupancies, the term "employees" targets people who staff building operational functions, such as management, operations, and maintenance personnel.

High-Rise Buildings. In the 2003 edition of the IFC, high-rise buildings were required to have fire and emergency evacuation plans, but drills were not required. In the 2006 code, requirements for fire and evacuation drills were added. Drills must now be conducted at least annually for the purpose of having occupants practice and become familiar with fire and evacuation procedures. As in the requirement for Group B discussed previously, the term "employees" in the high-rise building requirement is intended to encompass anyone who works in the building on a regular basis, including tenants, not just individuals who support building management, operations, or maintenance functions (see previous discussion for Group B).

Group R-2 College and University Buildings. With respect to Group R-2 college and university buildings, only occupancies that are within the scope of Table 405.2, Note d, are covered by Table 405.2. For guidance on how to interpret "Group R-2 college and university buildings," see the discussion on these occupancies under Section

Evacuation drill frequency and participation requirements for Group R-2 college and university buildings have been added to enhance safety in these occupancies.

404.2 herein. Group R-2 college and university buildings required to conduct fire and evacuation drills must do so at least four times per year, and all occupants must participate. The purpose of specifying four drills per year, as opposed to quarterly, is to recognize that occupant turnover follows a school's class schedule, which may not align with calendar quarters. The goal is to require a fire drill for each new group of residents who will inhabit the occupancy for a school session (semester, quarter, or similar class period).

408.3

Emergency Planning and Preparedness for Group E Occupancies and Group R-2 College and University Buildings

Fires in college and university housing often occur at night.

CHANGE TYPE. Addition

CHANGE SUMMARY. Requirements have been added detailing fire and evacuation drill procedures for Group R-2 college and university buildings. Correlating changes have been made to Section 404.2 and Table 405.2 to require fire and evacuation plans and periodic drills. **(F35-04/05)**

2006 CODE: 408.3 Group E Occupancies and Group R-2 College and University Buildings. Group E occupancies shall comply with the requirements of Sections 408.3.1 through 408.3.4 and Sections 401 through 406. Group R-2 college and university buildings shall comply with the requirements of Sections 408.3.1 and 408.3.3 and Sections 401 through 406.

408.3.3 Time of Day. Emergency evacuation drills shall be conducted at different hours of the day or evening, during the changing of classes, when the school is at assembly, during the recess or gymnastic periods, or during other times to avoid distinction between drills and actual fires. In Group R-2 college and university buildings one required drill shall be held during hours after sunset or before sunrise.

CHANGE SIGNIFICANCE. This change correlates with changes made to Section 404.2 and Table 405.2 in the 2006 edition, which require Group R-2 college and university buildings to have fire and emergency

This fraternity fire at the University of Mississippi claimed three lives. (Photo courtesy of Ed Comeau, Center for Campus Fire Safety.)

plans and periodic drills. Additional information on those changes is provided in the discussions for Section 402.4 and Table 405.2 herein.

The changes made to Section 408.3 provide details on when drills must be conducted. At least four drills must be conducted per year, as specified in Table 405.2. Section 408.3.1, which was not revised to specifically consider R-2 occupancies and correlate with this change, requires the first drill to be conducted within the first 10 days of the beginning of each school year. This section was originally written to apply to Group E occupancies. For Group R-2 college and university buildings, conducting drills within the first 10 days of each new school session, as opposed to each new school year, would be more prudent since student bodies may change significantly from one session to the next.

Section 408.3.3 requires at least one of the four annual drills to be conducted between sunset and sunrise. This requirement is intended to familiarize occupants with nighttime evacuation conditions, in view of the fact that many fires in residential occupancies occur during nighttime hours.

PART 3

Building and Site Requirements

Chapters 5 Through 10

- **Chapter 5**   Fire Service Features No changes addressed
- **Chapter 6**   Building Services and Systems
- **Chapter 7**   Fire-Resistance-Rated Construction No changes addressed
- **Chapter 8**   Interior Finish, Decorative Materials, and Furnishings
- **Chapter 9**   Fire Protection Systems
- **Chapter 10**   Means of Egress

This portion of the *International Fire Code* contains requirements that apply to all buildings and sites. Chapter 5 provides for fire service features, including provisions that provide firefighters with a means to access a building and establish a fire-protection water supply. Chapter 6 contains requirements for building systems such as elevators, standby and emergency power systems, stationary battery systems, and refrigeration systems. Buildings constructed using fire-resistive materials must be properly maintained to ensure that specified fire-resistance ratings are maintained, and Chapter 7 specifies the requirements for maintenance of fire-resistance rated construction.

Chapter 8, which was thoroughly revised in the 2006 edition, contains updated requirements for regulating interior finish materials, decorative materials, and furnishings.

Chapter 9, which is largely duplicated in the IBC, specifies the requirements for automatic sprinkler systems, alternative fire extinguishing systems, fire alarm and detection systems, standpipes, portable fire extinguishers, emergency alarm systems, and smoke control systems. For materials that can have a detonation or deflagration hazard, Chapter 9 specifies the requirements for explosion control systems.

Chapter 10, which is also largely duplicated in the IBC, establishes the minimum requirements for means of egress from buildings, including retroactive provisions applicable to all existing buildings, which are located at the back of the chapter. ∎

605.10
Portable Electric Space Heaters

606.8
Refrigerant Leak Detectors

606.9.1
Remote Controls for Refrigeration Systems

606.9.3
Manual Emergency Control Boxes for Refrigeration Systems

606.10
Emergency Pressure Control for Refrigeration Systems

608, 609, AND 602.1
Stationary Battery Systems

41

CHAPTER 8

Interior Finish, Decorative Materials, and Furniture

901.6.2

Records for Fire Protection Systems

901.9

Recall of Fire Protection Components

903

Automatic Sprinkler Systems—One- and Two-Family Dwellings and Townhouses

903.2.1.2

Group A-2—Automatic Sprinkler Systems

903.3.1.2.1

Balconies and Decks—Automatic Fire Sprinkler Installation Requirements

904.11.5.1

Portable Fire Extinguishers for Solid-Fuel Cooking Appliances

904.11.5.2

Class K Portable Fire Extinguishers for Deep-Fat Fryers

905.3.7

Standpipe Systems for Marinas and Boatyards

905.4

Location of Class I Standpipe Hose Connections at Horizontal Exits

906.2

Electronic Monitoring for Portable Fire Extinguishers

907.2.6

Fire Alarm Systems for Group I

907.2.9

Manual Fire Alarm Boxes in Group R-2

907.2.12.2

Emergency Voice/Alarm Communication Systems

907.10.1.2

Alarm Notification Appliances in Employee Work Areas

907.15

Monitoring of Fire Alarm and Detection Systems

910.1

Smoke and Heat Vents for ESFR Sprinklers

910.2.2

Smoke and Heat Vents for Group H

TABLE 910.3

Smoke Venting and Draft Curtains for High-Piled Storage

914

Fire Protection for Special Uses and Occupancies

1002.1

Definition of Accessible Means of Egress

1003.2

Minimum Ceiling Height

1003.3.2

Projection Limits on Freestanding Objects

1004.1

Determination of Design Occupant Load

TABLE 1004.1.1

Occupant Load Determination for Day Care Uses

1004.2

Maximum Occupant Load Permitted

1004.7

Occupant Load Determination for Fixed Seating

1007.1

Platform Lifts as Accessible Means of Egress

1007.3, 1007.4, 1007.6.2

Required Areas of Refuge

1008.1.1
Minimum Door Width in Group R-1 Occupancies

1008.1.2
Door Swing in Sleeping Units

1008.1.6
Thresholds at Residential Exterior Doors

1008.1.8.7
Remote Unlocking of Stairway Doors

1008.1.9
Panic and Fire Exit Hardware

1009.3.1, 1009.7
Curved Stairways

1009.5.2, 1010.7.2, 1014.5
Weather Protection of Exterior Egress Components

1009.5.3
Enclosed Usable Space under Stairways

1009.10, 1010.8, 1012
Handrails for Stairways and Ramps

1009.11.2, 1013.5, 1013.6
Protection at Roof-Hatch Openings

1010.6.3
Minimum Ramp Length

1010.9, 1010.9.1, 1010.9.2
Edge Protection at Ramps

1013.3
Guard Opening Limitations for Group
R-2 Occupancies

1014.2
Egress through Intervening Spaces

1014.2.1
Egress through Adjoining Tenant Spaces

1014.3
Common Path of Egress Travel in Group
R-2 Occupancies

1014.4.2
Aisle Accessways in Group M Occupancies

TABLE 1015.1
Single Means of Egress from Day Care Uses

1015.2.2
Egress Separation of Three or More Exits

1020.1
Unenclosed Interior Exit Stairways

1020.1.7.1
Egress from Smokeproof Enclosures

1025.3
Egress from Group A Occupancies

1028.2
Reliability—Maintenance of the Means of Egress

1028.4
Exit Signs—Maintenance

1028.7
Testing and Maintenance—Communication Systems
for Areas of Refuge

605.10

Portable Electric Space Heaters

Portable electric space heaters can cause a fire if they are used improperly.

CHANGE TYPE. Addition

CHANGE SUMMARY. This change provides minimum safety requirements for the use of portable electric space heaters in any occupancy. **(F54-04/05)**

2006 CODE: <u>**605.10 Portable Electric Space Heaters.** Portable electric space heaters shall comply with Sections 605.10.1 through 605.10.4.</u>

<u>**605.10.1 Listed and Labeled.** Only listed and labeled portable electric space heaters shall be used.</u>

<u>**605.10.2 Power Supply.** Portable electric space heaters shall be plugged directly into an approved receptacle.</u>

<u>**605.10.3 Extension Cords.** Portable electric space heaters shall not be plugged into extension cords.</u>

<u>**605.10.4 Prohibited Areas.** Portable electric space heaters shall not be operated within 3 feet (914 mm) of any combustible materials. Portable electric space heaters shall be operated only in locations for which they are listed.</u>

CHANGE SIGNIFICANCE. Regulations governing the selection and safe operation of portable electric space heaters, with the exception of provisions in Section 305.1, which require a safe clearance between sources of ignition and combustible materials have previously been overlooked by the IFC (Sections 603.4 and 603.5 only addressed fuel-fired heaters). The new Section 605.10 provides minimum requirements for the selection and use of portable electric space heaters to help ensure their safe operation.

By requiring that these devices be listed by a nationally recognized testing laboratory, appropriate safety features, such as protection against ignition due to tip-over and overheating of the heater itself, are required without specifically listing these detailed requirements in the IFC. The presence and functionality of these features are evaluated by the testing laboratory during the listing investigation. More details on minimum safety features can be found in the applicable Underwriters Laboratories listing standard UL 1278, "Standard for Movable and Wall- or Ceiling-Hung Electric Room Heaters."

Provisions in 605.10.2 and 605.10.3, which address power supply requirements, are intended to ensure that the power source is adequate to supply the unit without overloading and overheating the cord. It is preferable and consistent with the intent of the code change proponent to have portable heaters plugged directly into permanently mounted wall or floor receptacles, since the integral power cord will have been evaluated for suitability as part of the UL listing. Nevertheless, the code text does not literally mandate this. Although use of "extension cords" is specifically prohibited by the IFC, "power taps" (commonly referred to as power strips) listed to UL Standard

1363 are not "extension cords" and may technically be used if approved. Note that power taps are required to have integral overcurrent protection (see IFC Section 605.4.1), and it is important to make sure that the wattage rating of any power tap used meets or exceeds the wattage rating of the electric heater, which is required to be marked on the hang tag attached to the heater's power cord as a condition of the listing.

606.8

Refrigerant Leak Detectors

Refrigerant leak detector control panel.

CHANGE TYPE.   Modification

CHANGE SUMMARY.   The exception permitting omission of refrigerant leak detection equipment in ventilated ammonia refrigeration machinery rooms has been deleted. **(F58-04/05)**

2006 CODE: 606.8 Refrigerant Detector.   Machinery rooms shall contain a refrigerant detector with an audible and visual alarm. The detector, or a sampling tube that draws air to the detector, shall be located in an area where refrigerant from a leak will concentrate. The alarm shall be actuated at a value not greater than the corresponding TLV-TWA values shown in the *International Mechanical Code* for the refrigerant classification. Detectors and alarms shall be placed in approved locations.

> **Exception:** ~~Detectors are not required for ammonia systems where the machinery room complies with Section 1106.3 of the *International Mechanical Code*.~~

CHANGE SIGNIFICANCE.   This change resulted from a joint agreement between fire officials and the ammonia refrigeration industry to delete the longstanding allowance permitting ventilated ammonia machinery rooms to not have ammonia detectors. Several major players in the industry were surveyed to determine whether use of the current exception is commonplace, and there was general agreement that the exception was no longer needed. Most new machinery rooms are being provided with ammonia detection equipment to control the required emergency ventilation system rather than running the system continuously, which would otherwise be required without a leak detection system. Deletion of the exception has the effect of mandating ammonia detection in all ammonia refrigeration machinery rooms.

Refrigerant leak detector for ammonia.

606.9.1

Remote Controls for Refrigeration Systems

CHANGE TYPE.　Modification

CHANGE SUMMARY.　Emergency shutoff switches are no longer required for refrigeration machinery rooms containing only nonflammable refrigerants. **(F59-04/05)**

2006 CODE:　606.9.1 Refrigeration System.　A clearly identified switch of the break-glass type shall provide off-only control of electrically energized equipment and appliances in the machinery room, other than refrigerant leak detectors and machinery room ventilation.

> **Exception:** In machinery rooms where only nonflammable refrigerants are used, electrical equipment and appliances, other than compressors, are not required to be provided with a cut-off switch.

CHANGE SIGNIFICANCE.　An exception has been added to exclude refrigeration machinery rooms containing only nonflammable refrigerants from the requirement for an emergency electrical disconnect switch. The change recognizes that there is no ignition hazard associated with a release of a nonflammable refrigerant, such as a Freon gas.

In this case the code still requires a switch capable of shutting down compressors as an emergency response measure but not other electrically operated equipment that may be located in the machinery room.

Emergency control switches for refrigeration machinery rooms.

606.9.3

Manual Emergency Control Boxes for Refrigeration Systems

An inside view of an emergency control box for an ammonia refrigeration system.

CHANGE TYPE.   Deletion

CHANGE SUMMARY.   Manual emergency control boxes for refrigeration systems using toxic or flammable refrigerants or ammonia are no longer required. The replacement feature is an emergency pressure control system, now required by Section 606.10. **(F36-03/04)**

2006 CODE:

~~**606.9.3 Emergency Control Box.**~~   ~~Emergency control boxes shall be provided for refrigeration systems required to be equipped with a treatment system, flaring system or ammonia diffusion system.~~

~~**606.9.3.1 Location.**~~   ~~Emergency control boxes shall be located outside of the building at an approved accessible location. All portions of the emergency control box shall be 6 feet (1829 mm) or less above the adjoining grade.~~

~~**606.9.3.2 Construction.**~~   ~~Emergency control boxes shall be of iron or steel not less than 0.055 inch (1.4 mm) in thickness and provided with a hinged cover and lock.~~

~~**606.9.3.3 Operational Procedure.**~~   ~~Valves and switches shall be identified in an approved manner as to the sequential procedure to be followed in the event of an emergency.~~

A manual emergency control box for an ammonia refrigeration system.

606.9.3.4 Identification. ~~Emergency control boxes shall be provided with a permanent label on the outside cover reading: FIRE DEPARTMENT USE ONLY—REFRIGERANT CONTROL BOX, and including the name of the refrigerant in the system. Hazard identification in accordance with NFPA 704 shall be posted inside and outside of the control box.~~

606.9.3.5 Instructions. ~~Written instructions and information shall be provided and located in the emergency control box designating the following information:~~

> ~~1. Instructions for suspending operation of the system in the event of an emergency.~~
> ~~2. The name, address and emergency telephone numbers to obtain emergency service.~~
> ~~3. The location and operation of emergency discharge systems.~~

CHANGE SIGNIFICANCE. Enhancements in refrigeration system control equipment associated with new technologies now make it possible to provide an automatic emergency control system to replace key functions of the traditional manual emergency control box. Such automatic controls are now required by the new Section 606.10, which makes manual emergency control boxes obsolete. These control boxes were poorly understood and rarely utilized. Accordingly, Section 606.9.3, which required emergency control boxes in previous editions, has been deleted.

606.10

Emergency Pressure Control for Refrigeration Systems

A bank of compressors in a dated ammonia refrigeration machinery room.

CHANGE TYPE. Addition

CHANGE SUMMARY. Emergency pressure control systems are now required as a replacement for manual emergency control boxes for refrigeration systems utilizing toxic or flammable refrigerants or ammonia. Requirements for manual emergency control boxes, previously located in Section 606.9.3, are no longer required, and that section has been deleted from the code. **(F36-03/04)**

2006 CODE:

606.10 Emergency Pressure Control System. <u>Refrigeration systems containing more than 6.6 pounds (3 kg) of flammable, toxic or highly toxic refrigerant or ammonia shall be provided with an emergency pressure control system in accordance with Sections 606.10.1 and 606.10.2.</u>

606.10.1 Automatic Crossover Valves.
<u>Each high- and intermediate-pressure zone in a refrigeration system shall be provided with a single automatic valve providing a crossover connection to a lower pressure zone. Automatic crossover valves shall comply with Sections 606.10.1.1 through 606.10.1.3.</u>

606.10.1.1 Overpressure Limit Setpoint. <u>Automatic crossover valves shall be arranged to automatically relieve excess system pressure to a lower pressure zone if the pressure in a high- or</u>

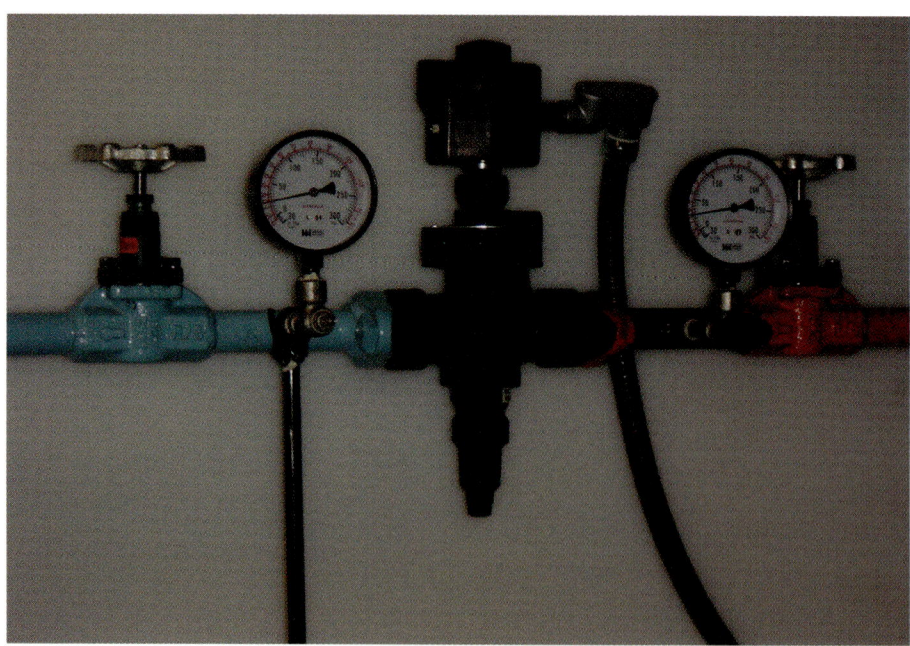

Emergency pressure control systems reduce the likelihood that pressure relief devices will discharge flammable, toxic, or highly toxic refrigerants or ammonia to the atmosphere. This is accomplished through the use of an automatic cross-over valve system that internally relieves excess pressure from high-pressure equipment to other portions of a refrigeration system before an atmospheric release occurs. This photo illustrates a "crossover" valve arrangement interconnecting the high-side with the low-side, which caused the pressure to equalize.

intermediate-pressure zone rises to within 15 psi (108.4 kPa) of the set point for emergency pressure-relief devices.

606.10.1.2 Manual Operation.
When required by the fire code official, automatic crossover valves shall be capable of manual operation.

606.10.1.3 System Design Pressure.
Refrigeration system zones that are connected to a higher pressure zone by an automatic crossover valve shall be designed to safely contain the maximum pressure that can be achieved by interconnection of the two zones.

606.10.2 Automatic Emergency Stop.
An automatic emergency stop feature shall be provided in accordance with Sections 606.10.2.1 and 606.10.2.2.

606.10.2.1 Operation of an Automatic Crossover Valve.
Operation of an automatic crossover valve shall cause all compressors on the affected system to immediately stop. Dedicated pressure-sensing devices located immediately adjacent to crossover valves shall be permitted as a means for determining operation of a valve.

To ensure that the automatic crossover valve system provides a redundant means of stopping compressors in an overpressure condition, high-pressure cutout sensors associated with compressors shall not be used as a basis for determining operation of a crossover valve.

606.10.2.2 Overpressure in Low-Pressure Zone.
The lowest pressure zone in a refrigeration system shall be provided with a dedicated means of determining a rise in system pressure to within 15 psi (103.4 kPa) of the setpoint for emergency pressure-relief devices. Activation of the overpressure sensing device shall cause all compressors on the effected system to immediately stop.

606.10 continues

Point of connection between the discharge pipe from a crossover valve (not shown) and a low-pressure receiver (tank).

606.10 continued

CHANGE SIGNIFICANCE. Enhancements in refrigeration system control equipment associated with new technologies have made it possible to provide an automatic emergency pressure control system (EPCS) to replace key functions of manual emergency control boxes, which were previously required by Section 606.9.3 for some refrigeration systems. The automatic controls required by the new Section 606.10 provide a means of mitigating an over-pressure condition prior to operation of emergency pressure-relief vents and, most likely, prior to the arrival of emergency responders. The automatic valves also eliminate the need for emergency responders to decipher the condition of a system in an attempt to determine whether operation of manual crossover valves in an emergency control box would be of benefit in mitigating a system malfunction.

The new provisions also ensure that lower pressure zones will be capable of handling additional pressure added by a crossover condition without over-pressurizing or operating the emergency relief vents on the lower zone. Prior codes did not address this concern, given the assumption that someone operating a manual crossover valve in the emergency control box would be knowledgeable with regard to system limitations, which may or may not be the case.

Nevertheless, the code never required the low-pressure side of the system to handle high-side pressure, and therefore some systems with emergency control boxes present the potential for an emergency responder to locally over-pressurize a system zone by fully opening a manual crossover valve too quickly. The resulting over-pressure condition could cause operation of a relief vent or even rupture the system.

Overall, the new Section 606.10 adds a requirement for a fully redundant safety control system in lieu of a manual system that has proven itself to be rarely, if ever, utilized by most fire departments. Elimination of manual controls favorably resolves long-standing industry concerns regarding the potential for harm caused by an untrained person operating valves in an emergency control box. There is no condition under which manual removal of refrigerant from a refrigeration system by the fire service is considered advisable. In contrast, automatic transfer of excess pressure to another zone of the system in conjunction with stopping the pressure source (compressors) can safely mitigate an over-pressure condition.

In the unlikely event that a fire causes the over-pressure condition, allowing system zones to automatically interconnect through the EPCS creates a much larger volume to limit pressure build-up while safely containing refrigerant. If the exposure fire continues to grow, emergency relief vents can protect the refrigeration system and automatic reseating valves can automatically limit the release of refrigerant to the amount necessary to maintain the system within design limits. In contrast, an emergency responder would not normally know how to properly cycle a manual valve in an emergency control box to limit the release of refrigerant to the minimum amount necessary for safety.

Key points regarding specific sections of the new code section are as follows:

606.10: The 6.6-pound threshold paralleled existing provisions in IFC Section 606.11.

606.10.1: The requirement for a single crossover valve between systems was based on the traditional industry practice of providing a single manual crossover valve in an emergency control box.

606.10.1.1: The 15-psi differential is intended to be a minimum range, and a larger differential is preferable if this can be provided, given operating and equipment constraints for a particular system. In many cases, relief valves can be expected to begin opening at pressures below the rated operating pressure, and a prudent design will take this into consideration by either increasing the differential between the EPCS set point and the rated activation pressure on a relief valve or specifically selecting a relief valve known to resist operation at less than its rated pressure.

606.10.1.2: The provision permits the local code official to require manual control capabilities for the crossover valve. Although this was not regarded as necessary from a safety perspective, it was recognized that some fire departments would be reluctant to completely give up manual controls.

606.10.2.1: The intent of this section is for the automatic crossover system to have a fully redundant means of stopping compressors. Compressors are ordinarily provided with automatic high-pressure cutout controls, but this section requires that these controls not be used to satisfy the new code requirement. An additional set of controls is required to serve as a back-up means of preventing a severe over-pressure condition that could cause operation of an emergency relief vent.

606.10.2.2: The lowest pressure zone of a system cannot be arranged to bleed pressure to another system zone, since crossing the lowest pressure zone to a higher pressure zone would most likely result in reverse flow. However, by providing a redundant emergency stop control, which would disengage the compressor, an over-pressure condition should be automatically mitigated. Over-pressure on a low-pressure zone would most likely result from a control valve that is stuck in the open position while transferring hot gas from the high-pressure side to defrost low-side components. In such a situation, stopping the compressor should disengage the pressure source for the defrost system. Note that compressors will cut out only if an over-pressure condition occurs. If an emergency condition involves a leak on the low side, compressors will continue to operate, which can be beneficial in pumping down the low side for this type of event.

To provide additional guidance on the design and installation of an EPCS, the International Institute of Ammonia Refrigeration has drafted a technical guideline which will most likely be published in the next edition of ANSI/IIAR 2, *Equipment, Design, and Installation of Ammonia Mechanical Refrigerating Systems.* Designers and code enforcers should refer to that document for further information. Note that a new section has been added to the *International Mechanical Code* (IMC), Section 1105.9, to reference the new IFC EPCS requirements.

608, 609, and 602.1

Stationary Battery Systems

CHANGE TYPE. Modification

CHANGE SUMMARY. Requirements for battery systems supplying standby, emergency, and uninterrupted power supplies have been updated to reflect current industry practices and to consolidate requirements into a single section. **(F39-03/04 and F61-04/05)**

2006 CODE: *For 2003 edition text of Sections 608 and 609, which were deleted and replaced by the following, see the 2003 IFC.*

~~**Section 608 Stationary Lead-Acid Battery Systems**~~ **Section 608 Stationary Storage Battery Systems**

608.1 Scope. Stationary storage battery systems having an electrolyte capacity of more than 50 gallons (189 L) for flooded lead acid, nickel cadmium (Ni-Cd) and valve-regulated lead acid (VRLA), or 1000 pounds (454 kg) for lithium-ion, used for facility standby power, emergency power or uninterrupted power supplies, shall comply with this section and Table 608.1.

TABLE 608.1 **Battery Requirements**

	Nonrecombinant Batteries		Recombinant Batteries	
Requirement	**Flooded Lead Acid Batteries**	**Flooded Nickel-Cadmium (Ni-Cd) Batteries**	**Valve Regulated Lead Acid (VLRA) Batteries**	**Lithium-Ion Batteries**
Safety caps	Venting caps (608.2.1)	Venting caps (608.2.1)	Self-resealing flame-arresting caps (608.2.2)	No caps
Thermal runaway management	Not required	Not required	Required (608.3)	Not required
Spill control	Required (608.5)	Required (608.5)	Required (608.3)	Not required
Neutralization	Required (608.5.1)	Required (608.5.1)	Required (608.5.2)	Not required
Ventilation	Required (608.6.1; 608.6.2)	Required (608.6.1; 608.6.2)	Required (608.6.1; 608.6.2)	Not required
Signage	Required (608.7)	Required (608.7)	Required (608.7)	Required (608.7)
Seismic protection	Required (608.8)	Required (608.8)	Required (608.8)	Required (608.8)
Smoke detection	Required (608.9)	Required (608.9)	Required (608.9)	Required (608.9)

608.2 Safety Caps. Safety caps for stationary storage battery systems shall comply with Sections 608.2.1 and 608.2.2.

608.2.1 Nonrecombinant Batteries. Vented lead acid, nickel-cadmium or other types of nonrecombinant batteries shall be provided with safety venting caps.

608.2.2 Recombinant Batteries. VRLA batteries shall be equipped with self-resealing flame-arresting safety vents.

608.3 Thermal Runaway. VRLA battery systems shall be provided with a listed device or other approved method to preclude, detect, and control thermal runaway.

608.4 Room Design and Construction. Enclosure of stationary battery systems shall comply with the *International Building Code.* Battery systems shall be allowed to be in the same room with the equipment they support.

608.4.1 Separate Rooms. When stationary batteries are installed in a separate equipment room accessible only to authorized personnel, they shall be permitted to be installed on an open rack for ease of maintenance.

608.4.2 Occupied Work Centers. When a system of VRLA, lithium-ion, or other type of sealed, nonventing batteries is situated in an occupied work center, it shall be allowed to be housed in a noncombustible cabinet or other enclosure to prevent access by unauthorized personnel.

608.4.3 Cabinets. When stationary batteries are contained in cabinets in occupied work centers, the cabinet enclosures shall be located within 10 feet (3048 mm) of the equipment that they support.

608.5 Spill Control and Neutralization. An approved method and materials for the control and neutralization of a spill of electrolyte shall be provided in areas containing lead-acid, nickel-cadmium or other types of batteries with free-flowing liquid electrolyte. For purposes of this paragraph, a "spill" is defined as any unintentional release of electrolyte.

> **Exception:** VRLA, lithium-ion, or other types of sealed batteries with immobilized electrolyte shall not require spill control.

608.5.1 Nonrecombinant Battery Neutralization. For battery systems containing lead-acid, nickel-cadmium, or other types of bat-

608, 609, and 602.1 continues

Battery systems used for standby, emergency, or uninterrupted power supplies are uniquely regulated by Section 608.

608, 609, and 602.1 continued

teries with free-flowing electrolyte, the method and materials shall be capable of neutralizing a spill from the largest lead-acid battery to a pH between 7.0 and 9.0.

608.5.2 Recombinant Battery Neutralization. For VRLA or other types of sealed batteries with immobilized electrolyte, the method and material shall be capable of neutralizing a spill of 3 percent of the capacity of the largest VRLA cell or block in the room to a pH between 7.0 and 9.0.

Exception: Lithium-ion batteries shall not require neutralization.

608.6 Ventilation. Ventilation of stationary storage battery systems shall comply with Sections 608.6.1 and 608.6.2.

608.6.1 Room Ventilation. Ventilation shall be provided in accordance with the *International Mechanical Code* and the following:

1. For flooded lead acid, flooded nickel-cadmium, and VRLA batteries, the ventilation system shall be designed to limit the maximum concentration of hydrogen to 1 percent of the total volume of the room; or

2. Continuous ventilation shall be provided at a rate of not less than 1 cubic foot per minute per square foot [1 ft^3/min/ft^2 or 0.0051 m^3/(s . m^2)] of floor area of the room.

Exception: Lithium-ion batteries shall not require ventilation.

608.6.2 Cabinet Ventilation. When VRLA batteries are installed inside a cabinet, the cabinet shall be approved for use in occupied spaces and shall be mechanically or naturally vented by one of the following methods:

1. The cabinet ventilation shall limit the maximum concentration of hydrogen to 1 percent of the total volume of the cabinet during the worst-case event of simultaneous "boost" charging of all the batteries in the cabinet; or

2. When calculations are not available to substantiate the ventilation rate, continuous ventilation shall be provided at a rate of not less than 1 cubic foot per minute per square foot [1 ft^3/min/ft^2 or 0.0051 m^3/(s .m^2)] of floor area covered by the cabinet. The room in which the cabinet is installed shall also be ventilated as required in Section 608.6.1.

608.7 Signage. Signs shall comply with Sections 608.7.1 and 608.7.2.

608.7.1 Equipment Room and Building Signage. Doors into electrical equipment rooms or buildings containing stationary battery systems shall be provided with approved signs. The signs shall state that:

1. The room contains energized battery systems.
2. The room contains energized electrical circuits.
3. The battery electrolyte solutions, where present, are corrosive liquids.

608.7.2 Cabinet Signage. Cabinets shall have exterior labels that identify the manufacturer and model number of the system and electrical rating (voltage and current) of the contained battery system. There shall be signs within the cabinet that indicate the relevant electrical, chemical, and fire hazards.

608.8 Seismic Protection. The battery systems shall be seismically braced in accordance with the *International Building Code.*

608.9 Smoke Detection. An approved automatic smoke detection system shall be installed in accordance with Section 907.2 in rooms containing stationary battery systems.

602.1 (Applicable Definitions)

STATIONARY STORAGE BATTERY, LEAD ACID. A group of electrochemical cells interconnected to supply a nominal voltage of DC power to a suitably connected electrical load, designed for service in a permanent location. The number of cells connected in series determines the nominal voltage rating of the battery. The size of the cells determines the discharge capacity of the entire battery. After discharge, it may be restored to a fully charged condition by an electric current flowing in a direction opposite to the flow of current when the battery is discharged.

NICKEL CADMIUM (Ni-Cd) BATTERY. An alkaline storage battery in which the positive active material is nickel oxide, the negative contains cadmium, and the electrolyte is potassium hydroxide.

NON-RECOMBINANT BATTERY. A storage battery in which, under conditions of normal use, hydrogen and oxygen gasses created by electrolysis are vented into the air outside of the battery.

RECOMBINANT BATTERY. A storage battery in which, under conditions of normal use, hydrogen and oxygen gases created by electrolysis are converted back into water inside the battery instead of venting into the air outside of the battery.

CHANGE SIGNIFICANCE. Section 608 provides unique requirements for battery systems used for standby, emergency, or uninterrupted power supplies. Because of these unique, hazard-specific re-

608, 609, and 602.1 continues

608, 609, and 602.1 continued

quirements, battery systems governed by Section 608 are exempted from the general requirements in the IFC for storage or use of corrosive liquids in Chapters 27 and 31, including fire sprinkler system and secondary containment requirements.

Section 608 in the 2006 IFC has been made more understandable through the elimination of unnecessary redundancy that existed between Sections 608 and 609 in the 2003 edition. Use of terminology that is consistent with industry standards also enhances usability. According to the proponent of these changes, all definitions related to Section 608 in the 2006 code (published in Section 602.1) were derived from official Institute of Electrical and Electronics Engineers (IEEE) Stationary Battery Committee documents.

The revised Section 608 also helps the IFC to better regulate several commonly used battery types that were not previously addressed, such as valve-regulated lead acid, nickel-cadmium (nicad), and lithium ion technologies. The regulations for lithium ion batteries are less stringent than those for other battery types because lithium ion technology is considered to be inherently safer. The proponent of this change indicated that lithium ion batteries used in back-up power systems are not prone to thermal runaway and are maintenance-free, making mitigation measures associated with thermal runaway and maintenance unnecessary. In addition, spill control and spill neutralization requirements are not applicable to lithium ion batteries because these batteries are not flooded with a liquid electrolyte, and ventilation requirements are not applicable to lithium ion batteries because there are no caps from which hydrogen gas could be released.

Chapter 8

Interior Finish, Decorative Materials, and Furniture

CHANGE TYPE. Modification

CHANGE SUMMARY. Chapter 8 was rewritten in its entirety to update and clarify provisions and to better recognize more current methodologies for fire-related material testing. **(F42-03/04, F43-03/04, F44-03/04, F47-03/04, F48-03/04, F49-03/04, F64-04/05, F67-04/05, F69-04/05, F70-04/05, F71-04/05, F72-04/05, F75-04/05, F76-04/05, F79-04/05, F80-04/05, F83-04/05, F84-04/05, F87-04/05, F91-04/05)**

2006 CODE: This chapter was revised in its entirety. For complete code text, see the 2006 *International Fire Code.*

CHANGE SIGNIFICANCE. Following the Station Nightclub fire in February 2003, in which 100 people were killed and more than 200 people were injured by a fire involving an improper interior finish material, IFC provisions for interior finish materials in Chapter 8 were subjected to a great deal of scrutiny. This, combined with other interest in updating material in Chapter 8 with regard to upholstered furniture and mattresses, resulted in a complete rewrite of the chapter in the 2006 edition. The new format, which was accomplished by Code Change F64–04/05, includes eight major sections, as follows:

801—General

802—Definitions

803—Interior Wall and Ceiling Finish and Trim in Existing Buildings

804—Interior Wall and Ceiling Trim in New and Existing Buildings

805—Upholstered Furniture and Mattresses in New and Existing Buildings

806—Decorative Vegetation in New and Existing Buildings

Chapter 8 continues

The use of expanded foam plastic as a ceiling finish is an example of an illegal interior finish that can result in rapid fire spread.

Chapter 8 continued

807—Decorative Materials Other Than Decorative Vegetation in New and Existing Buildings

808—Furnishings Other Than Upholstered Furniture and Mattresses or Decorative Materials in New and Existing Buildings

Using this section outline as a guide, the following discussion offers additional information on the changes that were made between the 2003 and 2006 code editions.

801.

This section was revised to reflect the expanded scope of Chapter 8 based on new requirements that have been added.

802.

No changes were made in this section. The section continues to reference Chapter 2 for definitions. See the new definition of "decorative materials" in Chapter 2.

803 (F91-04/05).

Most of the changes in Section 803, as compared to the 2003 code requirements, relate to improving organization and clarity; although, some technical changes were made. Section 803 was largely taken from Section 806 in the 2003 edition of the code. The other portion of Section 806 in the 2003 edition, which regulated interior trim, has been moved to Section 804 in 2006.

This section specifies the appropriate test methodologies and pass/fail criteria for interior finishes affixed to walls and ceilings. The title of the section includes "trim," based on material contained in Section 803.7.3, but there is no reason for this since Section 804 is devoted en-

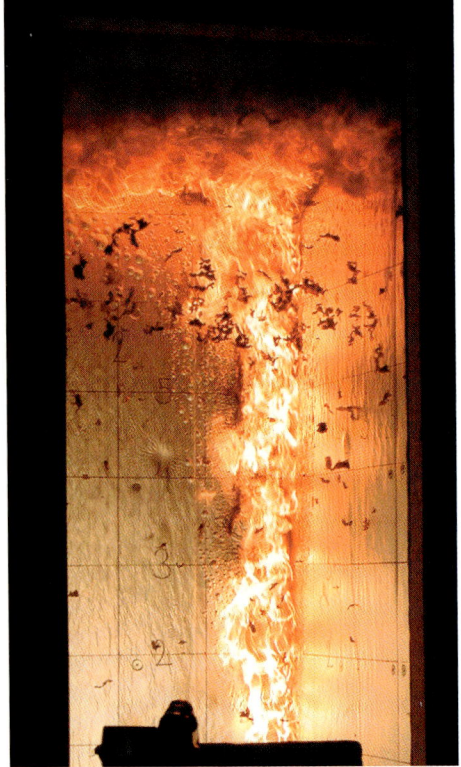

A view of the fire environment in a room corner test. This photo shows the room interior, looking through the door opening during a fire test of a wall covering material.

Room corner fire tests, conducted in an apparatus like this one, are recognized by the fire and building codes, and are a more representative method for evaluating fire performance of interior finish materials, as compared to the ASTM E84 tunnel test.

tirely to regulating trim and contains identical regulations. Duplication of trim regulations apparently resulted from multiple code changes that were involved in rewriting Chapter 8, which weren't fully coordinated.

Even though Chapter 8 has been updated in 2006, the code continues to recognize a minimum test methodology for wall and ceiling finishes that many consider to be outdated, ASTM E84 (also known at the Steiner tunnel test). Although the E84 test provides a reasonable basis for comparing surface flame-spread characteristics of "traditional" building materials, such as wood, the results of this test may not predict actual fire behavior for many contemporary materials, particularly textiles and plastics. Melting and dripping of these materials during the test and the use of a ceiling-mounted sample configuration can skew test results for contemporary materials, discrediting the E84 test results.

Because of these recognized weaknesses with the E84 test, the IFC references two more-current test methodologies, both classified as "room-corner tests," which are prescribed by NFPA Standards 265 and 286. Tests conducted in accordance with these standards are recognized by the IFC as alternatives to the ASTM E84 approach to flame-spread testing because room corner tests do a better job of simulating actual fire conditions. In a room corner test, materials are mounted to the walls and/or ceiling of a test room, and the fire exposure is generated by a gas burner that simulates a trash can fire extending to a chair in the corner of the room.

Of the two IFC-recognized room corner test methodologies, the NFPA 265 test is slightly less severe because the gas burner in this test does not operate at a high enough heat output to cause the burner flame to reach the ceiling. For this reason, recognition of the NFPA

Chapter 8 continues

The ASTM E84 Steiner tunnel is the traditional test apparatus for evaluating interior finish flame spread. This photo shows the test apparatus with the roof removed. Materials to be tested are mounted to the underside of the roof, and the roof is lowered onto the tunnel.

Chapter 8 continued   265 test is limited to evaluation of wall finishes only. In contrast, the burner flame in the NFPA 286 test contacts the ceiling surface in the latter portion of the test, providing a substantial fire exposure to ceiling-mounted materials. Because the burner flame exposes both the wall and the ceiling in the NFPA 286 test, this test can be used for evaluation of both wall and ceiling finishes.

The requirements of Section 803 can be summarized as follows.

1. Most interior wall and ceiling finish materials, with the exception of foam plastics, are permitted to be tested in accordance with either ASTM E84 or NFPA 286.

2. Textile wall finish materials are permitted to be tested in accordance with NFPA 286 or NFPA 265 or, if the building is sprinklered, a Class A flame spread index in accordance with ASTM E84 is permitted. The allowance to use NFPA 286 for textile finish materials is new to the 2006 edition. As indicated previously, NFPA 286 is a more severe test than NFPA 265, and if someone voluntarily chooses to use this method or employs it to accommodate use of a material as a ceiling finish, it is appropriate to recognize the results of the more severe test.

   Code users may wish to note that there are two methods of conducting the NFPA 265 test: Method A and Method B. Method A uses small samples of material whereas Method B requires covering large surface areas of the test room interior with material to be tested. The IFC recognizes both Method A and Method B, but the IBC recognizes only Method B. This difference between the codes resulted in part from differing opinions between the IFC and IBC code development committees that reviewed the respective proposals. Part of the reason for the difference of opinion between the IFC and IBC is based on the IFC's inclusion of regulations for both new and existing buildings, versus the IBC's addressing only new buildings. Given that many existing textile wall coverings in existing buildings were originally permitted to use Method A, mandating compliance with Method B for existing materials was regarded as being too restrictive for the IFC.

3. Textile ceiling finish materials are not addressed well by the IFC text. The intent of the IFC can be best derived by reviewing the IBC, which requires textile ceiling finish materials to pass the NFPA 286 test or, if the building is sprinklered, achieve a Class A flame-spread index in accordance with ASTM E84.

4. Expanded vinyl wall and ceiling finishes, which are typically thin vinyl materials with microscopic air bubbles creating a textured finish, are required to pass the NFPA 286 test or satisfy any of the testing options prescribed previously for textile finish materials.

5. Foam plastics are not permitted as interior wall or ceiling finishes except as allowed by IBC Section 2603, which prescribes unique and stringent regulations.

804.

This section addresses interior trim in both new and existing buildings. The 2006 requirements are derived from Sections 806.1.2 and 806.2.4 in the 2003 edition.

805 (F44-03/04, F45-03/04, F67-04/05, F69-04/05, F70-04/05, F71-05/05, F72-04/05, F75-04/05, F79-04/05),

GENERAL. This section addresses the use of upholstered furniture and mattresses in Group I-1, I-2, and I-3 occupancies with respect to smoldering ignition resistance and maximum heat release. Because numerous code changes simultaneously affected this section in the 2004-2005 code cycle and because of the complexity of the subject matter, some of the provisions in the 2006 edition will be further clarified in the 2006-2007 code cycle (which will affect the 2007 Supplement and the 2009 edition).

Based on exceptions provided throughout the section, most of these regulations are not applicable to furniture and mattresses in rooms and spaces equipped with fire sprinklers. Consequently, the regulations primarily impact existing facilities, since all new Group I occupancies are required by Chapter 9 to be equipped with a fire sprinkler system. An across-the-board elimination of the sprinkler exceptions for ignition-resistance and heat-release controls for upholstered furniture and mattresses had been proposed for Section 803, but the recommendation was not adopted by the ICC membership.

MATTRESS SMOLDERING IGNITION RESISTANCE.

With respect to mattresses, a technical change has been made to the 2006 edition to require smoldering ignition resistance in Group I-2 occupancies. A notable difference among the mattress regulations for Groups I-1, I-2, and I-3, which are now generally consistent with one another, is the lack of a sprinkler exception to the ignition-resistance requirement for mattresses in I-2. However, both this difference and the inclusion or exclusion of mattress regulations in the IFC are of minimal consequence, because all new mattresses in the United States are required to meet U.S. Consumer Product Safety Commission 16 CFR Part 1632, which prescribes a mandatory Federal smoldering ignition test. Consequently, the IFC's mattress ignition-resistance requirements have a limited impact, affecting only old mattresses (that preceded the Federal regulations) in existing Group I occupancies.

UPHOLSTERED FURNITURE IGNITION RESISTANCE.

Assessment of smoldering ignition resistance for upholstered furniture in Groups I-2 and I-3 is now permitted to be based on either NFPA 260 or NFPA 261, both of which are recognized tests. The omission of NFPA 261 as a recognized test for Group I-1 resulted from an oversight in the code-development process, which will likely be corrected in the 2007 Supplement.

NFPA 260, which is newly recognized by the IFC for testing of furniture in Group I-2 in 2006, is commonly referred to as the Upholstered Furniture Action Council (UFAC) test. IFC recognition of

Chapter 8 continues

this test was considered to be important because NFPA 260 is the most widely used test for evaluating smoldering ignition resistance of upholstered furniture. Use of either the NFPA 260 or 261 methodologies is considered satisfactory, as both tests provide a suitable basis of evaluation. The existence of two tests is attributed to the fact that different segments of the upholstered furniture industry have worked separately to develop their own test methodologies.

MATTRESS AND UPHOLSTERED FURNITURE COMBUSTIBILITY.

Additional changes to Section 805 in 2006 relate to limiting total energy and peak heat release rates for mattresses and upholstered furniture once they have been ignited. Requirements governing these combustion characteristics were made more restrictive in the 2006 edition for the purpose of coordinating the IFC values with values set by the California Bureau of Home Furnishings and Thermal Insulation in their Technical Bulletins #129 (Mattresses) and #133 (Furniture). As compared to the former IFC combustibility limits, the limits established by California's technical bulletins are believed to be more appropriate to accomplish the regulatory intent of controlling the contribution of mattresses and upholstered furniture to fire growth. The test methodologies prescribed by these California technical bulletins, which are widely used by industry and testing laboratories both inside and outside of California, were adopted by the IFC as equivalent to the already referenced ASTM E1590 and ASTM E1537 tests.

Another change to upholstered furniture regulations is the introduction of a new limit on heat release for furniture in Group I-1 occupancies. In the first and second printings of the 2006 edition, the total heat release limit is different from those for Groups I-2 and I-3, but this is an error in the code. The correct limit is 25 Megajoules in the first 10 minutes (not 25 MJ in the first 5 minutes), and this is also reflected in the IFC errata.

LABELING.

Finally, the 2006 edition establishes equivalent requirements for labeling upholstered furniture and mattresses in Groups I-1, I-2, and I-3. All upholstered furniture and mattresses in these occupancies must have a label indicating compliance with ignition-resistance and heat-release requirements set forth in the Section 805.

806 (F43-03/04, F84-04/05).

This section addresses decorative vegetation in new and existing buildings. The section is almost identical to Section 804 in the 2003 edition, with the changes being generally editorial in nature to improve the organization of requirements. The only significant technical change is the introduction of NFPA 701 as a basis for testing and approval of artificial vegetation, versus the 2003 edition requirement for materials to be flame resistant or flame retardant with no specific measurement criteria (see also Section 807 next).

807 (F42-03/04, F43-03/04, F47-03/04, F48-03/04, F49-03/04, F87-04/05).

This section addresses decorative materials, other than decorative vegetation. The section covers both new and existing buildings. The requirements are generally derived from Section 805 of the 2003 edition, but a significant modification has been made by referencing NFPA 701 as the basis for flammability testing and approval of nearly all decorative materials. In addition, a change has been made in Section 807.4.2.1, Exception 2 (formerly Section 803.2.1, Exception 2), to permit cellular and foam plastic trim up to 8 inches in width, versus 4 inches in the 2003 edition. This change correlates Section 807.4.2.1 with Section 803.7.3 (805.3.2 in 2003), which previously permitted sizes up to 8 inches.

FLAME RESISTANT AND FLAME RETARDANT VERSUS NFPA 701.

Previously, the code simply required materials to be "flame resistant" or "flame retardant," with no specific measurement criteria. These terms were not defined in the IFC, which led to difficulty in interpreting the code. The definitions should have been extracted from the Uniform Fire Code (UFC) when the IFC was originally written, because the UFC was the source for IFC provisions governing the flammability of decorative materials, but they were missed.

The UFC defined "flame retardant" to mean a chemical that can be applied to a material or fabric to render that material incapable of supporting combustion. In contrast, "flame resistant" was defined to mean materials that were treated with a flame retardant, were otherwise modified to resist ignition, or were inherently ignition resistant. If these definitions had been included in the IFC, then application of previous code editions would have been more straightforward, but without them, understanding the provisions proved challenging.

The 2006 edition's solution, which involved deleting the terms "flame resistant" and "flame retardant" and simply using NFPA 701 as a benchmark, solves problems in the previous edition by eliminating use of the undefined terms and by specifying a recognized basis for performance evaluation of materials.

DIFFERENTIATION OF WALL AND CEILING FINISHES AS COMPARED TO DECORATIVE MATERIALS.

A significant improvement in the 2006 edition is a new addition to Section 807.1, which clarifies the differences between wall and ceiling finishes versus decorative materials that are applied to walls and ceilings. It is important to point out that the third paragraph in Section 807.1, where this topic is handled, is not limited to any particular occupancy groups. Therefore, these regulations apply to all occupancies.

With the new third paragraph, the code now specifically states that materials placed on walls or ceilings after construction of the building must be regarded as interior finishes if they cover 10% or more of the wall or ceiling area to which they are applied. This is true regardless of the intended function of the material, be it structural, decorative,

Chapter 8 continues

Chapter 8 continued

acoustical, insulating, or otherwise. To be conservative, the 10% calculation should look at individual walls, as opposed to the aggregate of all walls in a room or space. Examples of materials encompassed in the 10% rule include paneling, wall pads, and crash pads.

The new third paragraph in Section 807.1 also captures finishes on fixed and movable walls and partitions, such as those used in modular office arrangements, and it requires finishes on both to comply with interior finish regulations, as opposed to regulations for decorations or furnishings.

The new fourth paragraph essentially serves as an exception to the fixed/movable wall and partition requirement in the third paragraph by allowing fabric partitions suspended from ceilings in Group B and Group M occupancies to be regulated as decorative materials because they more closely resemble draperies than interior finishes. Nevertheless, similar partitions suspended from the ceiling in an occupancy other than Group B or Group M would be required to meet the interior finish requirements because there is no exclusion for these other occupancies directing the code user to do otherwise.

Application of the code requirements for regulating the combustibility characteristics for fixed and movable walls and partitions in all occupancies can be summarized as follows:

1. The surface of a fixed or movable wall or partition that is supported from the floor must comply with the requirements for interior finishes.

2. The surface of a fixed or movable wall or partition that is suspended from the ceiling must comply with the requirements for interior finishes, unless the installation is in a Group B or Group M occupancy, in which case the materials may be tested in accordance with NFPA 701, as required for a decorative material.

When applying the code to suspended partitions, it is beneficial to consider the definition of "fabric partition" in IBC Section 1602 to capture the intent of the IFC provisions. The IBC definition was added as part of a companion code change that paralleled the Group B/Group M exception mentioned previously, and it conveys that a fabric partition consists of a fabric that is directly attached to a framing system with vertical framing members spaced greater than 4 feet on center and without a continuous rigid backing. An example would be a box frame with a piece of stretched fabric filling the inside of the frame. Uses for these units include providing a visual separation to subdivide a space or displaying advertising in a mercantile occupancy.

The intent of these new provisions is to help ensure that materials behaving as wall or ceiling surfaces, which have the potential to significantly impact room fire behavior, are tested to the more stringent protocols for interior finishes in ASTM E84, NFPA 265, and NFPA 286, as opposed to the less stringent decorative materials protocols in NFPA 701.

808 (F76-04/05, F80-04/05, F83-04/05).
The title of this section implies that the section addresses a variety of furnishings, other than upholstered furniture and mattresses and

decorative materials, in new and existing buildings. However, in the 2006 edition, only regulations for wastebaskets in detention facilities and plastic signs are included.

WASTEBASKETS IN GROUP I-3:

Wastebasket provisions were derived from Section 803.7.5 of the 2003 edition, but they have been substantially modified.

Provisions that are new to the 2006 edition include a limit on the peak rate of heat release for combustible containers and a requirement that large metal waste containers be listed in accordance with UL 1315. In contrast, the 2003 edition simply required waste containers to be noncombustible or constructed of "approved" combustible materials. Products evaluated in accordance with UL 1315 are evaluated to determine their ability to limit the external surface temperatures of the container bottom should their contents become ignited; extinguish a fire; and contain the contents without contributing fuel to a fire.

Summarizing the 2006 edition options, waste containers in Group I-3 are permitted to be constructed of either metallic or nonmetallic materials. If nonmetallic materials are used, the peak heat-release rate for the material is limited to 300 kW/m^2 when tested in accordance with ASTM E-1354. If metallic materials are used, the container must be listed in accordance with UL 1315 and have a noncombustible lid if the capacity is 20 gallons or more.

FOAM PLASTIC SIGNS.

Regulations for foam plastic signs, which have been added to the 2006 edition, are not entirely new. Similar requirements were included in the legacy fire codes (Uniform, BOCA, and Standard), but these weren't carried forward into the original IFC. Left unregulated, foam plastic signs could introduce a significant quantity of fast-burning fuel with a high heat-release potential into an occupancy. The new provisions limit the heat-release rate for foam plastic signs unless an exception is granted by the fire code official for signs not exceeding the lesser of 10% of the floor area or 10% of the wall area of the room or space where the sign is located.

901.6.2

Records for Fire Protection Systems

CHANGE TYPE. Modification (901.6.2) and Addition (901.6.2.1)

CHANGE SUMMARY. Mandatory records associated with fire protection systems have been expanded to require original installation records and to make installation and maintenance records more available to code officials for review. **(F97-04/05)**

2006 CODE: 901.6.2 Records. Records of all system inspections, tests, and maintenance required by the referenced standards shall be maintained on the premises for a minimum of 3 years and ~~made available~~ shall be copied to the fire code official upon request.

901.6.2.1 Records Information. Initial records shall include the name of the installation contractor, type of components installed, manufacturer of the components, location and number of components installed per floor. Records shall also include manufacturers' operation and maintenance instruction manuals. Such records shall be maintained on the premises.

CHANGE SIGNIFICANCE. Because different government agencies are sometimes involved in plan review and inspection of new versus existing occupancies, a jurisdiction's original installation records for fire protection systems may not be readily available to fire inspectors charged with inspection of existing occupancies. The change in the 2006 edition helps to ensure that installation records will remain available by requiring that copies be kept on site. A particular focus of this change was developing a mechanism to allow owners and inspectors the ability to readily identify installed fire protection equipment that has been recalled.

Building owners are now required to retain installation and maintenance records for fire protection systems on the premises.

   Although Section 901.6.2.1 is new to the IFC, the requirements in this section are not entirely new. Similar regulations have existed in several IFC reference standards for quite some time, including NFPA 72 for fire alarm equipment, NFPA 13 for fire sprinkler systems, and NFPA 25 for water-based fire protection systems. Rather than relying on provisions that are buried in reference standards, which may or may not be known to the owner, occupant, or fire official, the inclusion of a specific requirement in the IFC provides a concise regulation that is easily referenced.

   By ensuring that required records are available, current and future owners and occupants will benefit in addition to code enforcers. If design information is needed to evaluate the capabilities of an existing protection system to accommodate future changes in use or changes in stored commodities, it will be available to those who need it.

   The other revision to Section 901.6.2 authorizes the fire code official to require the owner to make copies of all inspection, test, and maintenance records for the jurisdiction's use, as opposed to the previous code requirement, which only mandated that the fire code official be given access to the owner's records.

901.9

Recall of Fire Protection Components

Recalled fire protection system components, such as this fire sprinkler, are required to be replaced with approved, listed components.

CHANGE TYPE.   Addition

CHANGE SUMMARY.   A section has been added to require the replacement of any recalled fire protection equipment and to require that the code official be notified when the replacement has been completed. **(F94-04/05)**

2006 CODE: <u>**901.9 Recall of Fire Protection Components.**</u> <u>Any fire protection system component regulated by this code that is the subject of a voluntary or mandatory recall under federal law shall be replaced with approved, listed components in compliance with the referenced standards of this code. The fire code official shall be notified in writing by the building owner when the recalled component parts have been replaced.</u>

CHANGE SIGNIFICANCE.   In recent years, a number of fire sprinklers and other fire-protection system components have been recalled, either under mandate by the U.S. Consumer Products Safety Commission (CPSC) or voluntarily by the manufacturer. The effectiveness of a recall in accomplishing replacement of recalled equipment varies according to a number of factors, including the ability of the CPSC or the manufacturer to notify affected customers and the willingness of customers to replace the equipment, which often involves some level of inconvenience.

The new provision in Section 901.9 provides a regulatory basis to require that recalled fire protection system components be replaced

with suitable equipment, and in doing so, it authorizes the fire code official to enforce the execution of a recall. In addition, the new provision requires that the owner notify the fire code official when replacement work has been completed so that this information can be entered into the jurisdiction's record-keeping system and perhaps be used as a basis for reporting the progress of a recall to CPSC and/or the affected manufacturer.

903

Automatic Sprinkler Systems—One- and Two-Family Dwellings and Townhouses

CHANGE TYPE. Addition

CHANGE SUMMARY. The 2006 *International Residential Code* (IRC) provides an adoptable appendix to require the installation of fire sprinklers in one- and two-family dwellings and town homes. Although this appendix is not duplicated in the IFC, fire officials should be aware of IRC Appendix P as an available basis for requiring sprinklering of one- and two-family dwellings and town homes. **(RB230-04/05)**

2006 CODE: *International Residential Code–Appendix P.* The provisions contained in this appendix are not mandatory unless specifically referenced in the adopting ordinance.

AP101 Fire Sprinklers. An approved automatic fire sprinkler system shall be installed in new one- and two-family dwellings and townhouses in accordance with 903.3.1 of the *International Building Code.*

CHANGE SIGNIFICANCE. This change is not published in the IFC, but it is of particular interest to fire code officials and has a bearing on enforcement of Section 903. Technically, as defined in Chapter 2 of the IFC, Group R occupancies are not intended to include one- and two-family dwellings and townhouses that are regulated by the IRC. Therefore, the requirement in IFC Section 903.2.7 for all buildings with a Group R fire area to be equipped with a fire sprinkler system is not intended to encompass one- and two-family dwellings or townhouses, and this interpretation of the code has been articulated by the International Code Council.

The new IRC Appendix P provides a concise means for jurisdictions to require fire sprinklers in one- and two-family dwellings and townhouses if such regulations are desired. The appendix is written in an adoptable format that can be enabled when the IRC is adopted.

2006 IRC Appendix P requires the installation of an approved automatic sprinkler system in one- and two-family dwellings and town homes, when adopted. With improvements in sprinkler technology, even affordable housing, such as this Habitat for Humanity home, can be sprinklered at minimal cost.

Given that the majority of fire deaths and injuries in the United States occur in residences, many code officials believe that requiring fire sprinklers in one- and two-family dwellings and townhouses is the ultimate solution to the nation's fire problem. There are many resources available to assist officials in making the case for residential sprinklers. These include the Home Fire Sprinkler Coalition (www.homefiresprinkler.org), the Residential Fire Safety Institute (www.firesafehome.org), and the International Association of Fire Chiefs booklet *Residential Fire Sprinklers. . . . A Step-by-Step Approach for Communities* (http://www.iafc.org/associations/4685/files/IAFCbooklet.pdf).

Today's residential sprinklers are designed with aesthetics in mind. Once installed, many types of residential sprinklers are barely noticeable.

Residential sprinkler systems are permitted to be installed using plastic piping, and can be combined with domestic plumbing service.

Sprinklers systems are an important component of residential fire safety since residential fires grow quickly, often creating life-threatening conditions in seconds.

903.2.1.2

Group A-2—Automatic Sprinkler Systems

CHANGE TYPE.   Modification

CHANGE SUMMARY.   The threshold for requiring fire sprinklers in Group A-2 occupancies has been reduced from 300 occupants to 100 occupants. **(F58-03/04)**

2006 CODE: 903.2.1.2 Group A-2.   An automatic sprinkler system shall be provided for Group A-2 occupancies where one of the following conditions exists:

1. The fire area exceeds 5000 square feet (464.5 m^2);
2. The fire area has an occupant load of ~~300~~ 100 or more; or
3. The fire area is located on a floor other than the level of exit discharge.

CHANGE SIGNIFICANCE.   Following the Station Nightclub fire in February 2003, in which 100 people were killed and more than 200 people were injured, model code regulations governing assembly occupancies were subjected to a great deal of scrutiny.

Even though the consequences of the Station fire were largely attributed to multiple code violations that were present on the night of the fire, particularly improper interior finish materials and the use of indoor pyrotechnics without proper precautions, the International Codes responded by changing the threshold for mandatory sprinklers in Group A-2 occupancies from 300 occupants to 100 occupants. The approach emphasizes the value of fire sprinklers as back-up protection for cases where other safety controls fail.

Although the Station fire involved a night club, the reduced sprinkler thresholds in the 2006 edition were extended to include all uses

Assembly occupancies intended for the consumption of food or drinks, such as this restaurant, will now require an approved automatic sprinkler system when the occupant load is 100 or more. (Photo courtesy of Denis Shell, Orange County, California, Fire Authority.)

classified as Group A-2 occupancies, which some people will regard as unjustified. Even restaurants that do not serve alcoholic beverages were impacted, but to the surprise of many, the restaurant industry did not oppose the change. Consequently, there was essentially no debate of the impact of reduced sprinkler thresholds on restaurants during the code-development process, and even fast food restaurants with large seating areas or an indoor playscape will require sprinklers because of this change.

The reduction of the sprinkler threshold for Group A-2 occupancies from 300 occupants to 100 occupants reflects a compromise that was reached among several competing proposals during the 2003–2004 code-development cycle. Some proposals recommended requiring fire sprinklers in all Group A occupancies, which essentially ties the requirement to an occupant load of 50, based on the definition of Group A, while others recommended tying the sprinkler threshold to consumption of alcoholic beverages or the presence of live entertainment.

In the end, the 100-occupant threshold was selected, in part because it correlated with the threshold for requiring panic hardware in Group A occupancies in Section 1008.1.9 of the 2003 edition (which was changed to 50 in 2006; see discussion of Section 1008.1.9 later). The other aspect of this decision was a belief by many that smaller assembly occupancies have a reduced life-safety risk because they can be readily evacuated.

903.3.1.2.1

Balconies and Decks— Automatic Fire Sprinkler Installation Requirements

A fire ignited by this barbecue grill caused significant damage.

CHANGE TYPE. Modification

CHANGE SUMMARY. This section now requires sprinklers for decks on some multifamily buildings, which were not technically encompassed by this section in the 2003 edition. **(F109-04/05)**

2006 CODE: 903.3.1.2.1 Balconies and Decks. Sprinkler protection shall be provided for exterior balconies, decks, and ground floor patios of dwelling units where the building is of Type V construction. Sidewall sprinklers that are used to protect such areas shall be permitted to be located such that their deflectors are within 1 inch (25 mm) to 6 inches (152 mm) below the structural members, and a maximum distance of 14 inches (356 mm) below the deck of the exterior balconies and decks that are constructed of open wood joist construction.

CHANGE SIGNIFICANCE. Section 903.3.1.2.1 was added to the IFC in 2003. The basis for that change was a recognition that the IBC grants many equivalencies (sometimes referred to as "trade-ups" or "trade-offs") based on the installation of NFPA 13R sprinkler systems, which do not ordinarily include sprinklers on balconies or decks because this level of protection is not required by NFPA 13R.

Recognizing that (1) balcony and deck fires account for a disproportionately large percentage of residential fire losses in multifamily buildings and (2) the IBC relies on NFPA 13R sprinkler systems for an enhanced level of property protection to justify code equivalencies, ICC members felt it necessary to overrule the NFPA 13R standard by adding Section 903.3.1.2.1 to require sprinklered balconies and decks.

The change in the 2006 edition clarifies the intended application of Section 903.3.1.2.1 by adding the term "decks" to the provisions.

Combustible decks and balconies associated with R-2 occupancies of Type V construction must be protected by automatic fire sprinklers.

Authors of the original text of Section 903.1.2.1 did not take into account that ICC codes differentiate between a "deck" and a "balcony" according to whether the building appendage is cantilevered (balcony) or supported at the far end (deck). Consequently, the term "decks" was inadvertently omitted when the section was originally developed.

Given that sprinklers were previously determined to be justified for protection of balconies, there was no reason for the code to have excluded decks from the sprinkler requirement simply because decks are supported differently than balconies.

904.11.5.1

Portable Fire Extinguishers for Solid-Fuel Cooking Appliances

Class K portable fire extinguishers, such as this one, are now required for the protection of small solid-fuel cooking appliances. (Photo courtesy of Amerex Corp.)

CHANGE TYPE.   Addition

CHANGE SUMMARY.   As a result of fire tests, Class K–rated portable fire extinguishers are now required for protecting small solid-fueled cooking appliances. **(F126-04/05)**

2006 CODE: <u>**904.11.5.1 Portable Fire Extinguishers for Solid Fuel Cooking Appliances.** All solid fuel cooking appliances, whether or not under a hood, with fireboxes 5 cubic feet (0.14 m³) or less in volume shall have a minimum 2.5 gallon (9 L) or two 1.5 gallon (6 L) K-rated wet chemical fire extinguishers located in accordance with Section 904.11.5.</u>

CHANGE SIGNIFICANCE.   Fires involving cooking appliances that use solid fuel can be difficult to extinguish, and the code previously lacked specific guidance on the size and type of fire extinguisher that is suitable for use with these appliances. This shortcoming has been addressed in the 2006 edition.

The burning of ordinary combustibles is classified as a Class A fire for the purpose of fire extinguisher selection. Accordingly, Class A–rated fire extinguishers have traditionally been chosen for protection of solid-fuel cooking appliances. However, with the introduction of Class K–rated extinguishers, a better choice is now available. Class K–rated fire extinguishers, which have become commonplace for protection of deep-fat fryers, have been found to be suitable for protection of all commercial cooking appliances.

In recognition of their unique effectiveness in extinguishing fires involving deep-fat fryers, Section 904.11.5.2 was added in 2006 to require Class K–rated extinguishers for protection of these appliances. That change will cause most commercial kitchens to require one or more Class K–rated extinguishers. Since these extinguishers will now be present in kitchens, it made sense to evaluate their capabilities with respect to extinguishing other types of kitchen fires.

Underwriters Laboratories tested the effectiveness of Class K extinguishers on small Class A fires, such as those found in solid-fuel appliances. Given that Class K extinguishing agents are water based (with water being recognized as an effective agent for Class A fires), it was no surprise that these tests were successful. Not only were fires suppressed, but the extinguishers were also found to provide enhanced operator safety because special spray nozzles control agent discharge.

Using the test results as justification, the ICC membership supported requiring Class K–rated fire extinguishers for protection of solid-fuel cooking appliances rather than simply offering them as a permissible alternative to Class A–rated extinguishers. A plus to recognition of Class K–rated extinguishers for multiple purposes is that many kitchens will now be able to utilize a single type of fire extinguisher for protection of both deep-fat fryers and solid-fuel cooking equipment, rather than needing multiple extinguisher types for protection of different cooking appliances.

To determine what agent quantity would constitute adequate protection with Class K extinguishers, the referenced fire tests were conducted using the UL 300 fire test protocol. These tests reportedly demonstrated that one 2.5-gallon (9-liter) or two 1.5-gallon (5.7-liter) K-rated extinguisher(s) will extinguish a fire in a 5-cubic-foot fire box. As a result, protection provided by this quantity of Class K agent (2.5 to 3 gallons) has been designated as equivalent to one 2-A rated portable fire extinguisher.

904.11.5.2

Class K Portable Fire Extinguishers for Deep-Fat Fryers

CHANGE TYPE. Addition

CHANGE SUMMARY. Guidance has been provided to assist with determining the required number and size of Class K–rated portable fire extinguishers for protection of deep-fat fryers. **(F127-04/05)**

2006 CODE: <u>**904.11.5.2 Class K Portable Fire Extinguishers for Deep-Fat Fryers.** When hazard areas include deep-fat fryers, listed Class K portable fire extinguishers shall be provided as follows:</u>

1. <u>For up to four fryers having a maximum cooking medium capacity of 80 pounds (36.3 kg) each: One Class K portable fire extinguisher of a minimum 1.5 gallon (6 L) capacity.</u>
2. <u>For every additional group of 4 fryers having a maximum cooking medium capacity of 80 pounds (36.3 kg) each: One additional Class K portable fire extinguisher of a minimum 1.5 gallon (6 L) capacity shall be provided.</u>
3. <u>For individual fryers exceeding 6 square feet (0.55 m^2) in surface area, Class K portable fire extinguishers shall be installed in accordance with the extinguisher manufacturer's recommendations.</u>

CHANGE SIGNIFICANCE. In the 2003 edition, Section 904.11.5 required Class K–rated fire extinguishers for protection of cooking equipment utilizing vegetable or animal oils and fats, but no guidance

Deep-fat fryers represent a significant challenge for fire extinguishing agents.

was provided with respect to the required number or size of these extinguishers to protect a particular hazard. Use of the previous provisions was also complicated because Class K extinguishers do not follow the traditional numerical rating system utilized by Class A- and B-rated fire extinguishers.

In the 2006 edition, guidance has been provided to assist end users and code officials in identifying an appropriate size and number of Class K–rated fire extinguishers needed for a particular kitchen, based on the number and size of deep-fat fryers present. The requirements set forth in Section 904.11.5.2 (outlined in the 2006 code text on the previous page) were recommended by fire equipment manufacturers on the basis of (1) input from end users and code officials and (2) the UL 300 test protocol utilizing a commercial fryer having an 80-pound capacity. The UL 300 test is the test that establishes Class K extinguisher ratings.

The discharge of a wet chemical fire extinguishing agent saponifies a layer on top of the burning liquid to help extinguish the fire.

905.3.7

Standpipe Systems for Marinas and Boatyards

CHANGE TYPE. Addition

CHANGE SUMMARY. A new section has been added to require standpipe systems for some marinas and boatyards, based on NFPA Standard 303. **(F99-03/04)**

2006 CODE: <u>905.3.7 Marinas and Boatyards.</u> <u>Marinas and boatyards shall be equipped throughout with standpipe systems in accordance with NFPA 303.</u>

CHANGE SIGNIFICANCE. The IBC and IFC have not previously included any special fire-protection-related requirements for protection of marinas or boatyards. Therefore, fire protection at these types of facilities has previously been relegated to local authorities, who will typically reference the NFPA standard on marina and boatyard fire protection, NFPA 303.

The new Section 905.3.7 and a companion modification to Chapter 45 change this approach by requiring, at a minimum, the installation of standpipe systems at many marinas and boatyards, using the requirements of NFPA 303 as a basis of regulation. The proponents of this change, a fire service group, justified the revision by pointing out that marinas and boatyards are typically large, expansive facilities that are often built over water. In many cases, these facilities have unique geographic features, making direct access by fire apparatus difficult and sometimes impossible. Standpipe systems help to offset challenges created by a lack of direct access by allowing the water supply for hose lines to be provided from a remote location.

Marinas and boatyards now require standpipe systems in accordance with NFPA 303. (Photo courtesy of South Kitsap, Washington Fire Rescue.)

Section 6.4.1 of NFPA 303 requires Class I standpipe systems, which provide 2.5-inch hose connections for fire department use, for piers, bulkheads, and buildings located more than 150 feet from fire apparatus. NFPA 303 requires that the distance must be determined by measuring along the path of hose lay as opposed to using a straight-line measurement. To determine the starting point for measuring the 150-foot distance, one must identify the closest location where fire apparatus access is provided. The IFC prescribes such a location to be the nearest point on a fire apparatus access road complying with Section 503.

The end point of the measurement is not clearly defined in NFPA 303. Under that standard, it could be interpreted to be either the nearest point of access to the pier (etc), or the most remote point. However, the IFC requires that the end point be the most remote point.

In the IFC, the basis for measuring distances for fire apparatus access layouts is set forth in Section 503.1.1. Section 503.1.1 specifies that access roadways must be arranged such that *all* portions of "facilities" must be within 150 feet of a fire apparatus access road, with Exception 2 allowing longer distances when alternative means of fire protection are provided. The term "facilities" is defined by the IFC to include "piers, wharves . . . and similar uses." Therefore, to comply with the apparatus access requirements in the IFC, all portions of a pier, etc., must be within 150 feet of a fire apparatus access road, or a standpipe system is necessary to gain longer distances based on Exception 2.

NFPA 303 permits marina and boatyard standpipe systems to be manual-dry systems, meaning that the only required water supply is a connection for fire apparatus and that piping is not required to contain water at all times. In addition, NFPA 303 permits the omission of hose racks, hoses, and standpipe cabinets at marinas and boatyards, in view of the fact that these would be likely targets for misuse by boat owners looking for a convenient hose for boat washing.

Fire department connection to a marina standpipe system. (Photo courtesy of South Kitsap, Washington, Fire Rescue.)

905.4

Location of Class I Standpipe Hose Connections at Horizontal Exits

A Class I standpipe hose connection is no longer required adjacent to a horizontal exit when the horizontal exit is within 130 feet of a hose connection located in an exit stairway.

CHANGE TYPE.   Modification

CHANGE SUMMARY.   This modification provides a limited exception to the requirement for standpipe hose connections to be installed at horizontal exits. **(F100-03/04)**

2006 CODE: 905.4 Location of Class I Standpipe Hose Connections.   Class I standpipe hose connections shall be provided in all of the following locations:

1. (No change to current text)
2. On each side of the wall adjacent to the exit opening of a horizontal exit.

Exception: Where floor areas adjacent to a horizontal exit are reachable from exit stairway hose connections by a 30-foot (9144 mm) hose stream from a nozzle attached to 100 feet (30,480 mm) of hose, a hose connection shall not be required at the horizontal exit.

3. through 6. (No change to current text.)

CHANGE SIGNIFICANCE.   Section 905.4, Item 2 has been revised to provide an exception to the requirement for installation of standpipe hose connections at horizontal exits where the distance between the horizontal exit and a hose connection in a stairway is less than 130 feet, as measured by hose-lay and nozzle trajectory distances. This exception previously existed in the BOCA code as Section 915.7, Exception 1, and it provides reasonable requirements for a design condition that is very common in buildings having horizontal exits located near stairways.

Building designs sometimes place stairways and horizontal exits in close proximity to one another. Additional hose connections at horizontal exits in these cases are of little to no value, since firefighters prefer to leave a stairway enclosure with a charged hose line when fighting a fire, as opposed to entering a floor in search of a standpipe hose connection. The 130-foot distance referenced in the exception was regarded as reasonable because Section 905.4(6) permits hose-lay distances of 150 to 200 feet before additional hose connections are possibly required for large-area buildings.

The proposal that led to this revision specifically contemplated situations where all floor areas on both sides of the horizontal exit are within the 130-foot distance from a stairway with a hose connection. In such cases, it is unnecessary to provide any hose connections at the horizontal exit.

However, there may be cases where floor areas on only one side of the horizontal exit are within the 130-foot distance, and the code text does not clearly address how to deal with such situations. Considering the logic used by the proponent to justify the new exception, it would be reasonable to permit omission of all standpipe connections at the horizontal exit based on the Exception to Item 2 in these situations, unless the distance limitations in Section 905.4 Item 6 are exceeded. Once the distance limitations in Item 6 are exceeded, additional hose connections at the horizontal exit would be appropriate. Note that hose connections, if provided to serve the floor area on only one side of the horizontal exit, should be placed on the opposite side of the separation wall so that firefighters will be able to access the connection from outside of the fire area that it serves.

906.2

Electonic Monitoring for Portable Fire Extinguishers

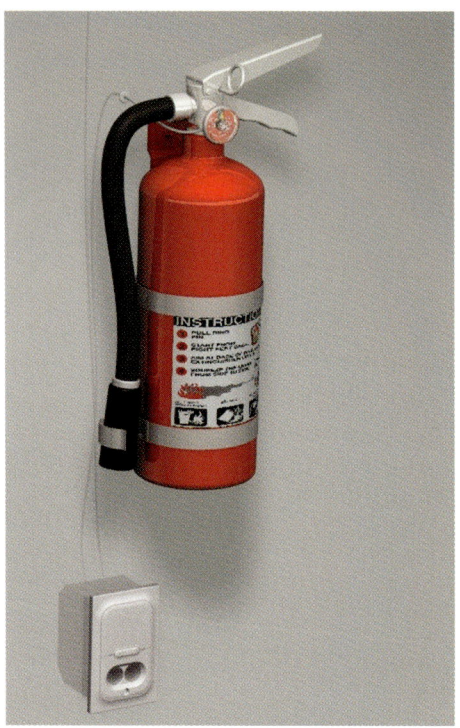

Listed and approved electronic monitoring equipment increases the reliability of portable fire extinguishers. (Photo courtesy of Mija Corporation.)

CHANGE TYPE. Addition

CHANGE SUMMARY. Electronic monitoring systems are now permitted for portable fire extinguishers in lieu of 30-day inspections and annual maintenance for up to 3 years. **(F104-03/04 and F135-04/05)**

2006 CODE: General Requirements. Fire extinguishers shall be selected, installed and maintained in accordance with this section and NFPA 10.

Exceptions:

1. The travel distance to reach an extinguisher shall not apply to the spectator seating portions of Group A-5 occupancies.

2. 30-day inspections shall not be required and maintenance shall be permitted to be once every three years for dry-chemical or halogenated agent portable fire extinguishers that are supervised by a listed and approved electronic monitoring device, provided:

    2.1. Electronic monitoring shall confirm that extinguishers are properly positioned, properly charged, and unobstructed, and

    2.2. Loss of power or circuit continuity to the electronic monitoring device shall initiate a trouble signal, and

    2.3. The extinguishers shall be installed inside of a building or cabinet in a non-corrosive environment, and

    2.4. Electronic monitoring devices and supervisory circuits shall be tested every three years when extinguisher maintenance is performed.

    2.5. A written log of required hydrostatic test dates for extinguishers shall be maintained by the owner to ensure that hydrostatic tests are conducted at the frequency required by NFPA 10.

CHANGE SIGNIFICANCE. Because the concept of using electronic monitoring devices to verify fire-extinguisher readiness is relatively new (Underwriters Laboratories only recently began listing devices for this purpose), NFPA 10 has not yet fully acknowledged the value of electronic monitoring technology. In contrast, the ICC membership has overwhelmingly embraced electronic monitoring, as demonstrated by nearly unanimous votes in support of the technology at the final hearings where it was considered.

The new Exception 2 to Section 906.2 serves to amend NFPA 10, which is otherwise referenced as the basis for regulating portable fire extinguishers in the IFC. The purpose of the exception is to permit the use of electronic fire extinguisher monitoring systems in lieu of 30-day inspections and annual maintenance, otherwise required by NFPA 10, for a period of up to 3 years for many fire extinguisher installations.

The traditional approach of NFPA 10 to inspection and maintenance of fire extinguishers allows significant intervals between times when extinguishers are checked. In contrast, electronic monitoring continuously ensures that extinguishers are properly mounted at required locations, are properly charged, and that there are no obstructions that would inhibit ready access by users. Any significant movement, tampering, or obstruction of an extinguisher; loss of pressure in the extinguisher; or loss of power in the monitoring equipment will generate a trouble signal on the supervisory alarm system, signaling the need for a physical inspection. Fire-extinguisher monitoring systems are also capable of maintaining an electronic log of extinguisher readiness and impairment history, which is useful to both owners and fire inspectors.

When this code change was discussed, a few code officials initially expressed concern that, without annual maintenance, agents in dry-chemical extinguishers might pack to the point that they would not discharge properly. However, testimony presented at the code hearing responded to that concern by indicating that lubricants added to modern dry-powder agents ensure proper discharge even when an extinguisher has remained in the same position for many years.

907.2.6

Fire Alarm Systems for Group I

CHANGE TYPE.   Clarification

CHANGE SUMMARY.   This change clarifies that a fire detection system is not required throughout all Group I-1 and I-2 occupancies, which was previously implied by Section 907.2.6. **(F111-03/04, F141-04/05)**

2006 CODE: **907.2.6 Group I.**   A manual fire alarm system ~~and an automatic fire detection system~~ shall be installed in Group I occupancies. An electrically supervised, automatic smoke detection system shall be provided ~~in waiting areas that are open to corridors~~ <u>in accordance with Sections 907.2.6.1 and 907.2.6.2</u>.

> **Exception:** Manual fire alarm boxes in <u>resident or</u> patient sleeping areas of Group I-1 and I-2 occupancies shall not be required at exits if located at all nurses' control stations or other constantly attended staff locations, provided such stations are visible and continuously accessible and that travel distances required in Section 907.4.1 are not exceeded.

<u>**907.2.6.1 Group I-1.**   Corridors, habitable spaces other than sleeping rooms and kitchens, and waiting areas that are open to corridors shall be equipped with an automatic smoke detection system.</u>

> <u>**Exceptions:**</u>
> 1. <u>Smoke detection in habitable spaces is not required where the facility is equipped throughout with an automatic sprinkler system.</u>
> 2. <u>Smoke detection is not required for exterior balconies.</u>

The requirements for fire detection systems in Group I-1 and I-2 occupancies have been clarified.

~~**907.2.6.1**~~ <u>**907.2.6.2**</u> **Group I-2.**   Corridors in nursing homes (both intermediate care and skilled-nursing facilities), detoxification facilities and spaces <u>permitted to be</u> open to the corridors <u>by IBC Section 407.2</u> shall be equipped with an automatic fire detection system. <u>Hospitals shall be equipped with smoke detection as required in Section 407.2 of the *International Building Code.*</u>

Exceptions:
1. and 2.   (No change)

CHANGE SIGNIFICANCE.   The code has been revised to clarify that fire detection systems are not required throughout all Group I-1 and I-2 occupancies, which was previously implied by Section 907.2.6. Instead, the code now provides specific direction on where fire detection is required. These changes reflect previously issued staff opinions on proper application of this section. In addition, the section was reformatted to split the requirements for fire detection in Groups I-1 and I-2 into two subsections, making them easier to identify.

Section 907.2.6.1, which covers Group I-1, calls for smoke detection to be provided in corridors; waiting areas that are open to corridors; and "habitable spaces" (defined in the IBC to include "spaces used for living, sleeping, eating or cooking other than bathrooms, toilet rooms, closets, halls, storage and utility spaces, and similar areas") other than sleeping units and kitchens. However, an exception largely negates the smoke detection requirement for habitable spaces because spaces equipped with automatic fire sprinklers are exempt, and all new Group I occupancies are required by Section 903.2.5 to be equipped with sprinklers.

907.2.6 continues

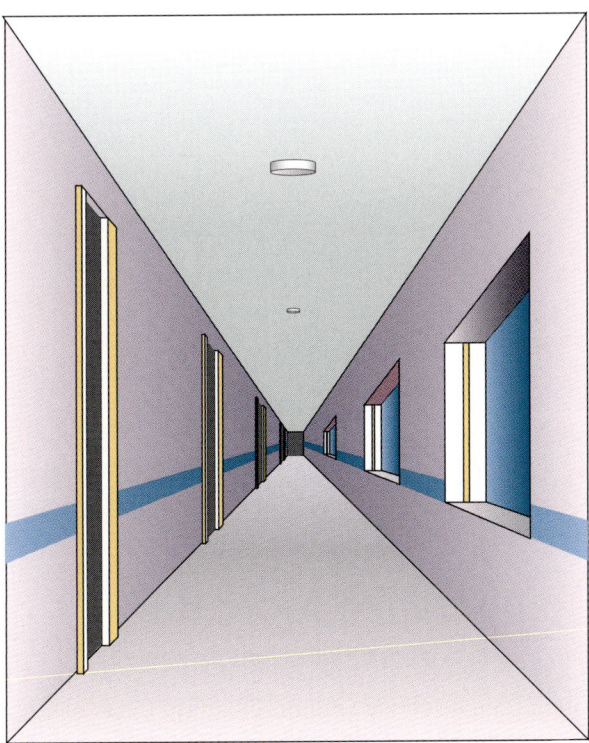

907.2.6 continued

Thus, the detection system requirements for Group I-1 in new construction essentially boil down to a requirement for smoke detectors in corridors and waiting areas open to corridors. Smoke detection is never required for kitchens in Group I-1 because of the potential for false alarms, and sleeping unit protection is excluded from this section in deference to smoke alarms required by Section 907.2.10. Smoke alarms (907.2.10) differ from smoke detectors (907.2.6) in that they are single-station or interconnected units without a control panel, as opposed to smoke detectors, which require a control panel to operate. Section 907.2.10.3 recognizes the use of a smoke detection system in lieu of individual smoke alarms if desired, but this is optional.

The new Section 907.6.2.2, which regulates Group I-2 occupancies, generally retains the text from the 2003 edition, except that a new sentence has been added to refer back to important requirements in IBC Section 407.2 for smoke detector placement in hospitals. IBC Section 407.2 generally prohibits corridors from being open to adjacent spaces but contains a number of allowances for waiting areas, nurses' stations, gift shops, and other special uses. Many of these allowances are conditional, based on detection being provided. Also note that IBC Section 407.6 provides additional detection requirements that should not be overlooked.

It should also be noted that the exception to Section 907.2.10.6 was updated by adding "resident" sleeping areas. The text of this exception previously mentioned only "patient" sleeping areas. Because some occupants in these facilities are not technically considered "patients," the addition of the term "resident" ensures that the exception can be properly applied as originally intended.

907.2.9

Manual Fire Alarm Boxes in Group R-2

CHANGE TYPE.   Modification

CHANGE SUMMARY.   This modification deletes the last remaining requirement for a manual fire alarm box in sprinklered Group R-2 occupancies. With this change, there are no longer any requirements for manual fire alarm boxes in Group R-2 occupancies because all such occupancies are required to be equipped with a fire-sprinkler system. **(F114-03/04)**

2006 CODE: 907.2.9 Group R-2.   A manual fire alarm system shall be installed in Group R-2 occupancies where:
   Items 1 through 3. (No change to current text)

Exceptions:
1. (No change to current text)
2. Manual fire alarm boxes are not required throughout the building when the following conditions are met:

    2.1. The building is equipped throughout with an automatic sprinkler system in accordance with Section 903.3.1.1 or Section 903.3.1.2; <u>and</u>
    2.2. The notification appliances will activate upon sprinkler flow ~~and~~.
    ~~**2.3.** At least one manual fire alarm box is installed at an approved location.~~

3. (No change to current text)

CHANGE SIGNIFICANCE.   Exception 2.3 has been deleted to correct an error that occurred during processing of Code Change F118-02, which resulted in the requirement for a single manual fire alarm box in Group R-1 uses to be unintentionally duplicated into the Group R-2 provisions. Although requiring a single manual fire alarm box makes sense in a hotel or motel, where it can be located behind the front desk, it did not make sense for apartment buildings because there is no consistent location where the device could be placed to be of any particular benefit.

Manual pull stations are not required in sprinklered Group R-2 occupancies when activation of the sprinkler system initiates an audible or visual alarm signal.

907.2.12.2

Emergency Voice/Alarm Communication Systems

CHANGE TYPE. Modification

CHANGE SUMMARY. This modification updates requirements for emergency voice/alarm communication system zoning. **(F120-03/04)**

2006 CODE: **907.2.12.2 Emergency Voice/Alarm Communication System.** The operation of any automatic fire detector, sprinkler water-flow device, or manual fire alarm box shall automatically sound an alert tone followed by voice instructions giving approved information and directions ~~on a general or selective basis to the following terminal areas~~ for a general or staged evacuation on a minimum of the alarming floor, the floor above, and the floor below in accordance with the building's fire safety and evacuation plans required by Section 404. Speakers shall be provided throughout the building by paging zones. As a minimum, paging zones shall be provided as follows:

1. Elevator groups ~~lobbies~~.
2. ~~Corridors~~ Exit stairways.

The requirements for emergency voice/alarm communication systems have been updated and clarified to be consistent with current design and installation practices for these systems.

3. ~~Rooms and tenant spaces exceeding 1000 square feet (93 m²)~~ ~~in area~~ <u>Each floor.</u>

4. ~~Dwelling units and sleeping units in Group R-2 occupancies.~~

5. ~~Sleeping units in Group R-1 occupancies.~~

~~6~~ <u>4.</u> Areas of refuge as defined in Section 1002.

Exception: In Group I-1 and I-2 occupancies, the alarm shall sound in a constantly attended area and a general occupant notification shall be broadcast over the overhead page.

907.2.12.2.1 Manual Override. A manual override for emergency voice communication shall be provided <u>on a selective and all-call basis</u> for all paging zones.

907.2.12.2.2 Live Voice Messages. The emergency voice/alarm communication system shall also have the capability to broadcast live voice messages through <u>paging zones on a selective and all-call basis</u> ~~speakers located in elevators, exit stairways, and throughout a selected floor or floors~~.

(no change to remaining text)

CHANGE SIGNIFICANCE. Revisions to this section update the requirements for voice/alarm communication systems, many of which can be traced back to the original adoption of voice paging system requirements in building codes in 1973. The previous text was a source of confusion as to whether Section 907.2.12.2 required speakers only in the listed areas or required speakers throughout, with separate paging zones for the listed areas. In either case, the requirements were inconsistent with modern approaches to designing and zoning voice/alarm communications systems for high-rise buildings. The 2006 revisions are intended to address these shortcomings by updating the requirements to be consistent with current practices.

907.10.1.2

Alarm Notification Appliances in Employee Work Areas

The design of the visual notification appliance circuits for new fire detection and alarm systems requires a minimum 20% spare capacity to accommodate additional devices that may be needed at a later date as a result of tenant improvements or building changes.

CHANGE TYPE. Modification

CHANGE SUMMARY. The modification provides an improved approach to accommodating the possible future need for visible-notification appliances for hearing-impaired employees. **(F153-04/05)**

2006 CODE: 907.10.1.2 Employee Work Areas. Where employee work areas have audible alarm coverage, the ~~wiring system shall be designed so that the visible alarm~~ notification appliance ~~s can be integrated into the alarm system.~~ circuits serving the employee work areas shall be initially designed with a minimum of 20% spare capacity to account for the potential of adding visible notification appliances in the future to accommodate hearing-impaired employee(s).

CHANGE SIGNIFICANCE. Provisions in the 2003 edition were not well stated and could have been interpreted to require a separate "spare" set of conductors at the time of original system installation to accommodate a possible future need for visible-notification appliances for hearing-impaired employees. The revision in the 2006 edition is intended to accommodate a reasonable number of additional visible appliances without having to install additional notification circuits. The approach is a much more cost-effective means of meeting the intent of this section, as compared to providing an additional set of conductors.

907.15

Monitoring of Fire Alarm and Detection Systems

CHANGE TYPE.   Modification

CHANGE SUMMARY.   All required fire alarm systems, including those that were previously provided only for sounding a local alarm, are now required to be monitored by an approved supervising station. (F155-04/05)

2006 CODE: 907.15 Monitoring.   ~~Where~~ Fire alarm systems required by this chapter or by the *International Building Code,* shall be monitored by an approved supervising station in accordance with NFPA 72~~shall monitor fire alarm systems~~.

> **Exceptions:** Supervisory service is not required for:
> 1. Single- and multiple-station smoke alarms required by Section 907.2.10.
> 2. Smoke detectors in Group I-3 occupancies.
> 3. Automatic sprinkler systems in one- and two-family dwellings.

CHANGE SIGNIFICANCE.   This section previously relied on other sections of the code to trigger a requirement for supervising station monitoring, but few other sections did this. The change in the 2006 edition provides for all code-required fire alarm systems to be monitored. By having a central station or proprietary supervising station monitor alarms, fire department response can be accelerated, which has a variety of benefits, including reduced losses and a reduced risk to firefighters by arriving on scene at an earlier stage of a fire event.

This code change is written so that alarm signals can be transmitted to either a central station or a proprietary supervising station. NFPA 72 recognizes either type of supervising station. The main dif-

907.15 continues

Required fire alarm systems must be monitored by an approved supervising station that complies with NFPA 72. (Photo courtesy of Property Protection Incorporated.)

907.15 continued

ference between the two is that a central station must be listed by an independent third-party organization, such as Factory Mutual or Underwriters Laboratories. A proprietary supervising station is not required to be listed.

Although the impact of this change initially appears to be very significant, it is moderated by the fact that most buildings with fire sprinkler systems are already required in Section 903.4 to have monitoring connections for sprinkler systems, and the threshold for requiring sprinklers in buildings is often lower than the threshold for requiring fire alarm systems.

910.1

Smoke and Heat Vents for ESFR Sprinklers

CHANGE TYPE.   Modification

CHANGE SUMMARY.   Automatic smoke and heat vents are no longer required for areas equipped with ESFR sprinklers. **(F126-03/04)**

2006 CODE: 910.1 General.   Where required by this code or otherwise installed, smoke and heat vents, or mechanical smoke exhaust systems and draft curtains shall conform to the requirements of this section.

> **Exceptions:**
> 1. Frozen food warehouses used solely for storage of Class I and Class II commodities where protected by an approved sprinkler system.
> 2. Where areas of buildings are equipped with early suppression fast-response (ESFR) sprinklers, automatic smoke and heat vents shall not be required within these areas.

CHANGE SIGNIFICANCE.   Requirements to provide smoke and heat vents in sprinklered buildings have been controversial for decades, but they continue to remain in the IFC, having been carried over from the Uniform and Standard fire codes. A common belief among many fire protection professionals is that automatic smoke and heat vents may interfere with the activation and/or effectiveness of sprinklers if the vents open before sprinklers activate. Many also believe that, even if sprinklers do activate first, smoke and heat vents will not provide the intended benefit because sprinkler discharge cools smoke, which in turn reduces smoke buoyancy and vent effectiveness.

Supporters of smoke and heat vents counter that vents are important as an aide to firefighters when sprinklers do not control a fire. In such cases, smoke and heat vents may reduce or eliminate the need for firefighters to manually vent the roof using saws to facilitate interior fire-attack strategies. Supporters also argue that vents are beneficial during salvage and overhaul operations by providing a means to remove residual smoke from a building.

910.1 continues

Areas of buildings protected by Early Suppression Fast Response (ESFR) sprinklers no longer require automatic smoke and heat vents.

A row of smoke and heat vents on the roof of a warehouse building.

910.1 continued

With the advent of ESFR sprinklers, the smoke and heat vent debate took on a new dimension. Unlike traditional "control mode" sprinkler design criteria, which were often based on extrapolation and/or engineering judgment, design criteria for ESFR sprinklers are based on extensive full-scale fire test programs, none of which involved smoke and heat venting. Results of these tests serve as the basis for ESFR design criteria, and the design criteria leave little room for unanticipated variables that might lead to activation of additional sprinklers. In contrast, control mode designs are, at least in theory, better able to accommodate the uncertainty of sprinkler-vent interaction because control mode designs anticipate a much larger number of operating sprinklers to control a fire.

To better quantify the effect of smoke and heat vents on ESFR sprinkler performance, fire tests were eventually conducted, and according to the proponent of this code change, these tests demonstrated the potential for inadequate performance by ESFR sprinklers when automatic smoke and heat vents are present. Of particular concern is the risk of delayed sprinkler activation due to vent operation, which could result in a fire overwhelming ESFR sprinklers. This concern ultimately convinced ICC members to approve the new Exception 2 to Section 910.1, which deletes requirements for automatic smoke and heat vents where ESFR sprinklers are used.

This change improves internal consistency among various code sections that require smoke and heat vents and draft curtains, including Section 910.1, Section 910.3.5 (Exception 1), which permits omission of draft curtains where ESFR sprinklers are installed, and Footnote "j" to Table 2306.2, which permits omission of smoke and heat vents, mechanical smoke removal systems, and draft curtains in ESFR-protected high-piled combustible storage areas. The ESFR-related allowances in Section 910.3.5 and Table 2306.2 were both approved in the 2003 edition for reasons similar to those outlined previously.

One other aspect of this section worth noting is that manually operated vents can still be required in large Group F-1 and S-1 occupancies, based on Section 910.2.1, even though automatic vents are not required. This is true because Section 910.1 Exception 2 only excepts the requirement for automatic vents and because Section 910.3.2 requires vents to be both automatically and manually operable. Applying the literal text of the code, the manual vent requirement could still be considered by some to be applicable. Nevertheless, having manual vents without draft curtains would be of questionable value in large buildings since draft curtains are needed to build smoke depth below vents, which helps them to perform effectively. Furthermore, in view of the fact that high-piled storage provisions in Chapter 23 completely exempt ESFR-protected areas from smoke and heat venting requirements, the logic of requiring manual vents in large F-1 and S-1 occupancies without high-piled storage would be questionable.

CHANGE TYPE. Modification

CHANGE SUMMARY. Requirements for smoke and heat vents based exclusively on classification of an occupancy as Group H have been deleted. **(F132-03/04 [Deleted Item 1] and F162-04/05 [Deleted Item 2])**

2006 CODE: ~~910.2.2 Group H. Buildings and portions thereof used as a Group occupancy as follows:~~

~~1. In occupancies classified as Group H-2 or H-3, any of which are more than 15,000 square feet (1394 m²) in single floor area.~~

~~**Exception:** Buildings of noncombustible construction containing only noncombustible materials.~~

~~2. In areas of buildings in Group H used for storing Class 2, 3 and 4 liquid and solid oxidizers, Class 1 and unclassified detonable organic peroxides, Class 3 and 4 unstable (reactive) materials, or Class 2 or 3 water reactive materials as required for a high-hazard commodity classification.~~

~~**Exception:** Buildings of noncombustible construction containing only noncombustible materials.~~

CHANGE SIGNIFICANCE. This revision is actually the result of two code changes, one of which deleted Item 1, and one of which deleted Item 2.

Item 1 was deleted because it was no longer considered necessary. The requirements in Item 1 originated at least 25 years ago in the Uniform Building Code based on speculation that smoke and heat vents would be of value in reducing fire damage and improving inte-

910.2.2 continues

910.2.2

Smoke and Heat Vents for Group H

Group H occupancies, such as this one, no longer require automatic smoke and heat vents based solely on the Group H classification.

910.2.2 continued rior conditions during a fire. However, a lack of evidence to support this hypothesis for sprinklered buildings and enhancements to Group H occupancy fire- and life-safety features made over the years caused many to question the validity of the requirement and ultimately led to its deletion.

Item 2 was deleted because no technical basis could be identified to support the original requirement and because the value of providing smoke and heat venting based exclusively on the presence of small quantities of hazardous materials was not evident. The previous IFC required any Group H room, no matter how small, to be provided with smoke and heat venting. Thereby, the provision for smoke and heat venting was triggered even when a room contained only 1 to 2 pounds of some materials, such as detonable organic peroxides and Class 4 oxidizers, or 2 to 4 gallons of other materials, such as Class 3 water reactives and Class 3 oxidizers.

Given that smoke and heat vents are generally intended as an aide to firefighters, theoretically making interior attack less challenging by venting the fire, these vents provide no apparent benefit in occupancies containing highly reactive or explosive materials because interior fire attack in such occupancies would be ill-advised once a fire becomes large enough to operate automatic vents.

The effort to delete special Group H venting requirements was bolstered by the fact that the IBC and IFC lacked any associated design criteria, so even when vents were required by the old codes, designers had no identifiable basis for vent sizing or spacing.

CHANGE TYPE.   Modification and Clarification

CHANGE SUMMARY. In addition to editorial clarifications, the Table 910.3 has been modified to specify design criteria for smoke and heat vents where draft curtains are not provided. **(F134-03/04 and F135-03/04)**

Table 910.3

Smoke Venting and Draft Curtains for High-Piled Storage

2006 CODE:

TABLE 910.3   Requirements for Draft Curtains and Smoke Vents[a]

Occupancy Group and Commodity Classification	Designated Storage Height (feet)	Minimum Draft Curtain Depth (feet)	Maximum Area Formed by Draft Curtains (square feet)	Vent Area to Floor Area Ratio[c]	Maximum Spacing of Vent Centers (feet)	Maximum Distance to Vents from Wall or Draft Curtains[b] (feet)
Groups F-1 and S-1	—	$0.2 \times H^d$ but >4	50,000	1:100	120	60
~~Group S-1~~ High-piled Storage (See Section 910.2.3)-I-IV (Option 1)	≤20	6	10,000	1:100	100	60
~~Group S-1~~ High-piled Storage (See Section 910.2.3) I-IV (Option 2)	>20 ≤40	6	8000	1:75	100	55
~~Group S-1~~ High-piled Storage (See Section 910.2.3) High hazard (Option 1)	≤20	4	3000	1:75	100	55
~~Group S-1~~ High-piled Storage (See Section 910.2.3) High hazard (Option 2)	>20 ≤40	4	3000	1:50	100	50

For SI: 1 foot = 304.8 mm, 1 square foot = 0.0929m².

a. Requirements for rack storage heights in excess of those indicated shall be in accordance with Chapter 23. For solid-piled storage heights in excess of those indicated, an approved engineered design shall be used.

b. The distance specified is the maximum distance from any vent in a particular draft curtained area to walls or draft curtains which form the perimeter of the draft curtained area.

c. Where draft curtains are not required, the vent area to floor area ratio shall be calculated based on a minimum draft curtain depth of 6 feet (Option 1).

d. "H" is the height of the vent, in feet, above the floor.

Table 910.3 continues

Table 910.3 continued

CHANGE SIGNIFICANCE.   This revision fills several gaps in previous code editions related to smoke and heat vents and draft curtains. The first fix is accomplished by the new Footnote "c" to Table 910.3, and it relates to high-piled storage areas having smoke and heat vents but not draft curtains. Table 2306.2 requires this level of protection for:

1. high-piled storage areas without ESFR protection that exceed 12,000 square feet in area and contain Class 1 through 4 commodities and
2. high-piled storage areas without ESFR protection that exceed 2500 square feet in area and contain "high hazard" commodities.

Chapter 23 always references Section 910 for smoke and heat vent design criteria, but for areas without draft curtains, Section 910 previously provided no guidance. The new Footnote "c" to Table 910.3 resolves this issue, as least in part, by directing that the vent-to-floor-area ratio be calculated on the basis of an assumed draft curtain depth of 6 feet. What is not stated, but is implied, is that the maximum center-to-center spacing of vents and the maximum distance from perimeter walls to the nearest vent should be consistent with the selected vent-to-floor-area ratio.

The second fix is accomplished by the new Footnote "d" to Table 910.3. This footnote was added as an editorial clarification to specify that the "H" factor used in calculating draft curtain depth for F-1 and S-1 occupancies relates to the height of the vent, in feet, above the floor. The change is based on guidance provided by NFPA Standard 204, which is a comprehensive design standard for smoke and heat venting and draft curtains.

Several changes have been made to the code to clarify requirements for draft curtains and smoke and heat vents.

The third fix has been accomplished by revising the first column to Table 910.3 to segregate smoke and heat vent and draft curtain requirements that are occupancy-based from those that are based on high-piled combustible storage. High-piled storage can exist in several occupancies, including Group M "big box" stores, and the previous version of Table 910.3 addressed only Group S-1 high-piled storage areas. Now, the table correlates with the triggers in Section 910.2 by having (1) a row for F-1 and S-1 occupancies that require vents and draft curtains based on floor areas and (2) several additional rows that address high-piled combustible storage, regardless of occupancy.

914

Fire Protection for Special Uses and Occupancies

The IBC fire protection requirements for special uses and occupancies such as atriums, malls, and high-rise buildings have been duplicated into IFC Section 914. (Photo courtesy of Orange County, California Fire Authority.)

CHANGE TYPE. Addition

CHANGE SUMMARY. Requirements for fire protection based on special uses and occupancies have been duplicated into the IFC from the IBC so that the IFC requirements for fire protection systems are complete. **(F140-03/04)**

2006 CODE:

914.1 General. This section shall specify where fire protection systems are required based on the detailed requirements of use and occupancy of the *International Building Code.*

914.2 Covered Mall Buildings. Covered mall buildings shall comply with Sections 914.2.1 through 914.2.4.

914.2.1 Automatic Sprinkler System. The covered mall building and buildings connected shall be equipped throughout with an automatic sprinkler system in accordance with Section 903.1.1, which shall comply with the following:

1. The automatic sprinkler system shall be complete and operative throughout occupied space in the covered mall building prior to occupancy of any of the tenant spaces. Unoccupied tenant spaces shall be similarly protected unless provided with approved alternate protection.
2. Sprinkler protection for the mall shall be independent from that provided for tenant spaces or anchors. Where tenant spaces are supplied by the same system, they shall be independently controlled.

Exception: An automatic sprinkler system shall not be required in spaces or areas of open parking garages constructed in accordance with Section 406.2 of the *International Building Code.*

914.2.2 Standpipe System. The covered mall building shall be equipped throughout with a standpipe system in accordance with Section 905.

914.2.3 Emergency Voice/Alarm Communication System.
Covered mall buildings exceeding 50,000 square feet (4645 m) in total floor area shall be provided with an emergency voice/alarm communication system. Emergency voice/alarm communication systems serving a mall, required or otherwise, shall be accessible to the fire department.

The system shall be provided in accordance with Section 907.2.12.2.

914.2.4 Fire Department Access to Equipment. Rooms or areas containing controls for air-conditioning systems, automatic fire-extinguishing systems or other detection, suppression, or control elements shall be identified for use by the fire department.

914.3 High-Rise Buildings.   High-rise buildings shall comply with Sections 914.3.1 through 914.3.5.

914.3.1 Automatic Sprinkler System.   Buildings and structures shall be equipped throughout with an automatic sprinkler system in accordance with Section 903.3.1.1 and a secondary water supply where required by Section 903.3.5.2.

>   **Exceptions:** An automatic sprinkler system shall not be required in spaces or areas of:
>
>   1. Open parking garages in accordance with Section 406.3 of the *International Building Code.*
>   2. Telecommunication equipment buildings used exclusively for telecommunications equipment, associated electrical power distribution equipment, batteries and standby engines, provided that those spaces or areas are equipped throughout with an automatic fire detection system in accordance with Section 907.2 and are separated from the remainder of the building with fire barriers consisting of not less than 1-hour fire-resistance-rated walls and 2-hour fire-resistance-rated floor/ceiling assemblies.

914.3.2 Automatic Fire Detection.   Smoke detection shall be provided in accordance with Section 907.2.12.1.

914.3.3 Emergency Voice/Alarm Communication System.   An emergency voice/alarm communication system shall be provided in accordance with Section 907.2.12.2.

914 continues

Section 914.4 establishes special fire-protection requirements for atriums.

Section 914.2 establishes special fire-protection requirements for malls. (Photo courtesy of Orange County, California Fire Authority.)

914 continued

914.3.4 Fire Department Communication System. A two-way fire department communication system shall be provided for fire department use in accordance with Section 907.2.12.3.

914.3.5 Fire Command. A fire command center complying with Section 509 shall be provided in a location approved by the fire department.

914.4 Atriums. Atriums shall comply with Sections 914.4.1 and 914.4.2.

914.4.1 Automatic Sprinkler System. An approved automatic sprinkler system shall be installed throughout the entire building.

> **Exceptions:**
> 1. That area of a building adjacent to or above the atrium need not be sprinklered, provided that portion of the building is separated from the atrium portion by not less than a 2-hour fire barrier.
> 2. Where the ceilings of the atrium are more than 55 feet (16,764 mm) above the floor, sprinkler protection at the ceiling of the atrium is not required.

914.4.2 Fire Alarm System. A fire alarm system shall be provided where required by Section 907.2.13.

914.5 Underground Buildings. Underground buildings shall comply with Sections 914.5.1 through 914.5.6.

914.5.1 Automatic Sprinkler System. The highest level of exit discharge serving the underground portions of the building and all levels below shall be equipped with an automatic sprinkler system installed in accordance with Section 903.3.1.1. Water-flow switches and control valves shall be supervised in accordance with Section 903.4.

914.5.2 Smoke Control System. A smoke control system is required to control the migration of products of combustion in accordance with Section 909 and provisions of this section. Smoke control shall restrict movement of smoke to the general area of fire origin and maintain means of egress in a usable condition.

914.5.3 Compartment Smoke Control System. Where compartmentation is required by Section 405.4 of the *International Building Code,* each compartment shall have an independent smoke-control system. The system shall be automatically activated and capable of manual operation in accordance with Section 907.2.18.

914.5.4 Fire Alarm System. A fire alarm system shall be provided where required by Section 907.2.19.

914.5.5 Public Address. A public address system shall be provided where required by Section 907.2.19.1.

914.5.6 Standpipe System. The underground building shall be provided throughout with a standpipe system in accordance with Section 905.

914.6 Stages. Stages shall comply with Sections 914.6.1 and 914.6.2.

914.6.1 Automatic Sprinkler System. Stages shall be equipped with an automatic fire-extinguishing system in accordance with Chapter 9. Sprinklers shall be installed under the roof and gridiron and under all catwalks and galleries over the stage. Sprinklers shall be installed in dressing rooms, performer lounges, shops and storerooms accessory to such stages.

Exceptions:
1. Sprinklers are not required under stage areas less than 4 feet (1219 mm) in clear height utilized exclusively for storage of tables and chairs, provided the concealed space is separated from the adjacent spaces by not less than $\frac{5}{8}$-inch (15.9 mm) Type X gypsum board.
2. Sprinklers are not required for stages 1000 square feet ($93m^2$) or less in area and 50 feet (15,240 mm) or less in height where curtains, scenery, or other combustible hangings are not retractable vertically. Combustible hangings shall be limited to a single main curtain, borders, legs, and a single backdrop.
3. Sprinklers are not required within portable orchestra enclosures on stages.

914.6.2 Standpipe System. Standpipe systems shall be provided in accordance with Section 905.

914.7 Special Amusement Buildings. Special amusement buildings shall comply with Sections 914.7.1 and 914.7.2.

914.7.1 Automatic Sprinkler System. Special amusement buildings shall be equipped throughout with an automatic sprinkler system in accordance with Section 903.3.1.1. Where the special amusement building is temporary, the sprinkler water supply shall be of an approved temporary means.

Exception: Automatic sprinklers are not required where the total floor area of a temporary special amusement building is less than 1000 square feet ($93 m^2$) and the travel distance from any point to an exit is less than 50 feet (15,240 mm).

914.7.2 Automatic Fire Detection. Special amusement buildings shall be equipped with an automatic fire detection system in accordance with Section 907.2.11.

914 continues

914 continued

914.8 Aircraft-Related Occupancies. Aircraft-related occupancies shall comply with Sections 914.8.1 through 914.8.5.

914.8.1 Automatic fire detection systems. Airport traffic control towers shall be provided with an automatic fire detection system installed in accordance with Section 907.2.

914.8.2 Fire Suppression. Aircraft hangars shall be provided with fire suppression as required by NFPA 409.

> **Exception:** Group II hangars, as defined in NFPA 409, storing private aircraft without major maintenance or overhaul are exempt from foam suppression requirements.

914.8.3 Finishing. The process of "doping," involving the use of a volatile flammable solvent, or of painting shall be carried on in a separate detached building equipped with automatic fire-extinguishing equipment in accordance with Section 903.

914.8.4 Residential Aircraft Hangar Smoke Alarms. Smoke alarms shall be provided within residential aircraft hangars in accordance with Section 907.2.21.

914.8.5 Aircraft Paint Hangar Fire Suppression. Aircraft paint hangars shall be provided with fire suppression as required by NFPA 409.

914.9 Application of Flammable Finishes. An automatic fire-extinguishing system shall be provided in all spray, dip, and immersing spaces and storage rooms, and shall be installed in accordance with Chapter 9.

914.10 Drying Rooms. Drying rooms designed for high-hazard materials and processes, including special occupancies as provided for in Chapter 4 of the *International Building Code,* shall be protected by an approved automatic fire-extinguishing system complying with the provisions of Chapter 9.

CHANGE SIGNIFICANCE. This change adds a new Section 914 to duplicate fire-protection system requirements related to special uses and occupancies from the IBC into the IFC. The source of this material is IBC Chapter 4. Special uses and occupancies addressed by this section include:

1. covered mall buildings
2. high-rise buildings
3. atriums
4. underground buildings
5. stages
6. special amusement buildings

7. airport traffic control towers and hangars

8. areas used for application of flammable finishes

9. areas used for drying of goods with flammable finishes.

Having a single section in IFC Chapter 9 to reflect all of the special use and occupancy fire protection provisions in IBC Chapter 4, generally in the same order, was considered more user-friendly than distributing these requirements into the various sections of IFC Chapter 9 that deal with sprinklers, alarms, etc. Nevertheless, many of these requirements were already reflected elsewhere in Chapter 9, and further work will be required to correlate the overlapping provisions.

In the meantime, code users should be careful not to overlook some important sprinkler, standpipe, alarm, and smoke control-related provisions in Section 914 that are not duplicated in other parts of Chapter 9 (where they would otherwise be expected based on the type of system). These include, among others, an exception to sprinkler requirements for high-rise telecommunications areas in Section 914.3.1 and requirements for sprinklers and alarms in both temporary and permanent special amusement buildings, none of which appear in Sections 903 or 907, where one would expect to find them.

1002.1

Definition of Accessible Means of Egress

CHANGE TYPE. Clarification

CHANGE SUMMARY. The definition of an accessible means of egress has been modified to indicate that it begins at those areas of a building required to be accessible and ends at a public way. **(E6-03/14; E75-04)**

2006 CODE: 1002.1 Definitions. Accessible Means of Egress. A continuous and unobstructed way of egress travel from any <u>accessible</u> point in a building or facility ~~that provides an accessible route to an area of refuge, a horizontal exit or~~ <u>to</u> a public way.

CHANGE SIGNIFICANCE. The previous definition seemed to indicate that an accessible means of egress terminated at an area of refuge or a horizontal exit. This modified definition clarifies that an accessible means of egress will include any exit stairway or elevator that occurs beyond an area of refuge or horizontal exit and does not end until reaching the public way. Some persons may need assistance utilizing the exit and exit discharge, and safe evacuation to a public way must be addressed for all building occupants.

The code change also coordinates with Section 1007.1 in that an accessible means of egress is not mandated from those points in a building that are not required to be accessible. Where a space or floor level is not required to be accessible, there is no reason to provide an accessible means of egress.

An accessible means of egress.

CHANGE TYPE.   Modification

CHANGE SUMMARY.   The minimum ceiling height throughout the means-of-egress system has been increased from 7 feet to 7 feet 6 inches. **(E8-03/04)**

2006 CODE: 1003.2 Ceiling Height.   The means of egress shall have a ceiling height of not less than 7 feet 6 inches (2134 mm) (2286 mm).

Exceptions: (no change to text)

CHANGE SIGNIFICANCE.   Section 1208.2 requires nearly all occupiable spaces to have a minimum ceiling height of 7 feet 6 inches. While most of the occupiable floor area of a building must be considered a means of egress due to the almost unlimited potential paths of egress travel, the provisions of Section 1208.2 are broader in scope so as to make the requirement in Section 1003.2 have limited application. The modified ceiling height requirement now provides for consistency between the two sections.

The increase in the minimum ceiling height has no effect on the reductions in height permitted by Exceptions 1 through 5. For example, Exception 3 still allows protruding objects such as light fixtures, sprinklers, and signs to extend below the minimum required height. Such projections must provide for a minimum headroom clearance of 80 inches, and the reduction is limited to 50% of the ceiling area. The change to provide for a minimum ceiling height of 7 feet 6 inches does not modify the required 80-inch clearance.

1003.2
Minimum Ceiling Height

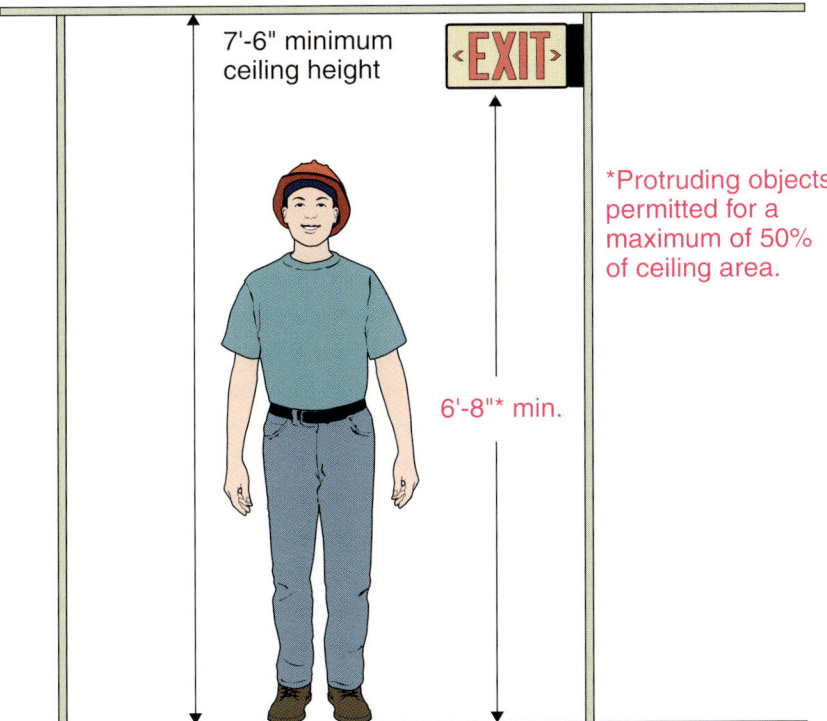

7'-6" minimum ceiling height

<EXIT>

*Protruding objects permitted for a maximum of 50% of ceiling area.

6'-8"* min.

Minimum Ceiling Height in a Means of Egress Has Been Increased

1003.3.2

Projection Limits on Freestanding Objects

CHANGE TYPE.   Modification

CHANGE SUMMARY.   The permitted overhang of a freestanding object mounted on a post or pylon has been reduced from 12 inches to 4 inches where the lowest point of the leading edge is more than 27 inches and less than 80 inches above the walking surface. **(E3-04/05)**

2006 CODE: 1003.3.2 Freestanding Objects.   A freestanding object mounted on a post or pylon shall not overhang that post or pylon more than ~~12~~ 4 inches (~~305~~ 102 mm) where the lowest point of the leading edge is more than 27 inches (686 mm) and less than 80 inches (2032 mm) above the walking surface. Where a sign or other obstruction is mounted between posts or pylons and the clear distance between the posts or pylons is greater than 12 inches (305 mm), the lowest edge of such sign or obstruction shall be 27 inches (685 mm) maximum or 80 inches (2032 mm) minimum above the finish floor or ground.

> **Exception:** This requirement shall not apply to sloping portions of handrails serving stairs and ramps.

CHANGE SIGNIFICANCE.   Where an object mounted on a post or pylon protrudes into an egress path, it becomes a potential obstruction, particularly to those individuals who are sight-impaired. The code has previously permitted an overhang of up to 12 inches beyond the post or pylon where the lowest point of the leading edge of the object is located more than 27 inches and less than 80 inches above the walking surface. The allowable overhang has now been reduced to 4 inches, making the limitation consistent with the accessibility provisions of ICC A117.1 Section 307.3.

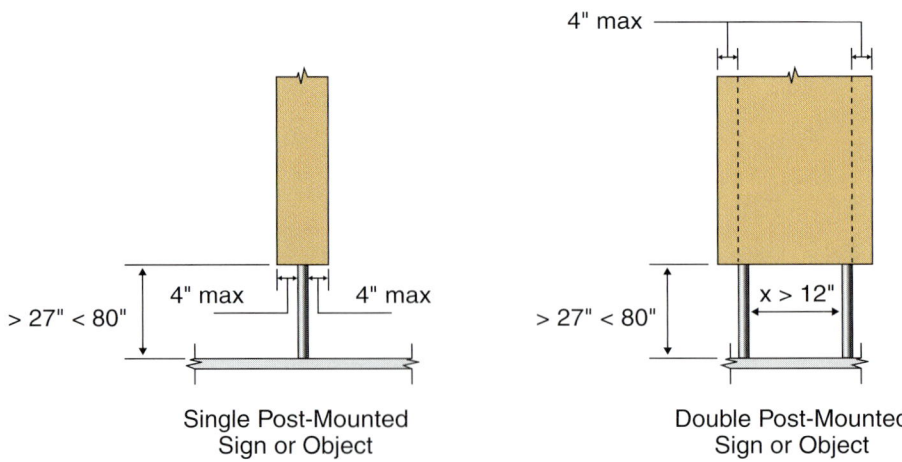

Single Post-Mounted
Sign or Object

Double Post-Mounted
Sign or Object

Projection Limits on Freestanding Objects Extending Up from the Floor

The limitations on overhang are of primary importance to those individuals with visual impairments. Initial research indicates that numerous body contacts are likely to be made with objects that protrude 12 inches from a post or pylon, even by adults who travel using a long cane proficiently with one of the three principal cane techniques. Persons of shorter stature, children, and those persons who do not consistently use proficient techniques with a long cane are particularly at risk. The reduction in the permitted overhang beyond a post or pylon significantly reduces the potential for accidental impacts.

1004.1

Determination of Design Occupant Load

CHANGE TYPE. Addition

CHANGE SUMMARY. A new exception permits the building official to assign an occupant load to a room, space, or building that is less than that calculated with Table 1004.1.1. Previously, no allowance was available for reducing a calculated occupant load. In addition, the revised text more clearly describes the procedure for determination of the design occupant load. Additional text now sets forth a means for the establishment of the "occupant per unit of area" factor where the intended use is not listed in Table 1004.1.1. **(E9-03/04)**

2006 CODE: 1004.1 Design Occupant Load. In determining means of egress requirements, the number of occupants for whom means of egress facilities shall be provided shall be ~~established by the largest number computed~~ determined in accordance with ~~Sections 1004.1.1 through 1004.1.3~~ this section. Where occupants from accessory areas egress through a primary space, the calculated occupant load for the primary space shall include the total occupant load of the

Example:
Given: A 210,000 square foot industrial building designed for final assembly of commercial aircraft.

Determine: The design occupant load of the building.

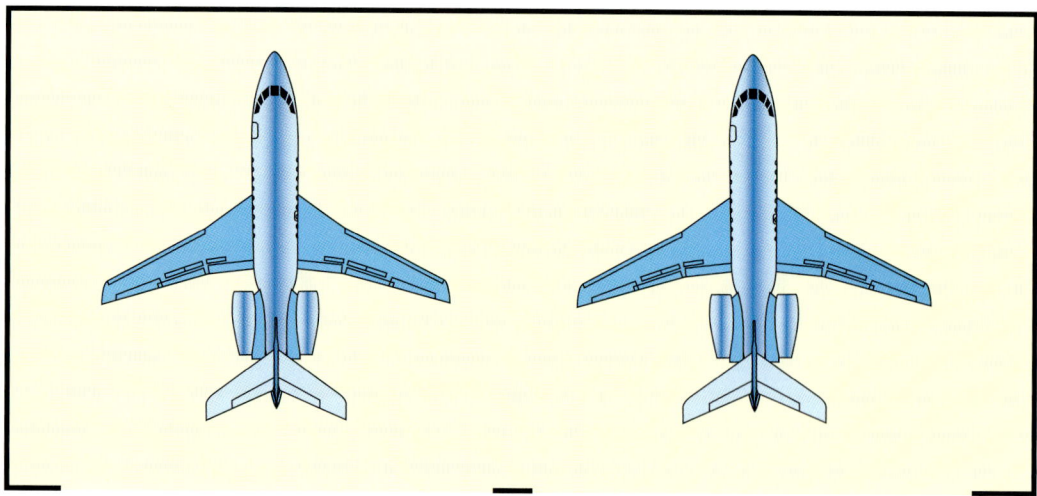

Based upon Table 1004.1.1, the design occupant load would be 2100.

$$\left(\frac{210{,}000 \text{ square feet}}{100 \text{ (factor for industrial areas)}} \right)$$

Where approved by the building official, a more realistic design occupant load is permitted based on the actual maximum number of occupants anticipated in the building.

Determination of Design Occupant Load

primary space plus the number of occupants egressing through it from the accessory area.

1004.1.1 Actual Number. ~~The actual number occupants for whom each occupied space, floor or building is designed.~~

1004.1.~~2~~.1 ~~Number by Table 1004.1.2.~~ **Areas without Fixed Seating.** The number of occupants shall be computed at the rate of one occupant per unit of area as prescribed in Table 1004.1.~~2~~ .1. For areas without fixed seating, the occupant load shall not be less than that number determined by dividing the floor area under consideration by the occupant per unit of area factor assigned to the occupancy as set forth in Table 1004.1.1. Where an intended use is not listed in Table 1004.1.1, the building official shall establish a use based on a listed use that most nearly resembles the intended use.

> **Exception:** Where approved by the building official, the actual number of occupants for whom each occupied space, floor, or building is designed, although less than those determined by calculation, shall be permitted to be used in the determination of the design occupant load.

1004.1.3 Number by Combination. ~~Where occupants from accessory spaces egress through a primary area, the calculated occupant load for the primary space shall include the total occupant load of the primary space plus the number of occupants egressing through it from the accessory space.~~

CHANGE SIGNIFICANCE. The new exception allows for a reduction in the calculated design occupant load on a very limited case-by-case basis. Previously, the ability to utilize the actual number of occupants was available only where the anticipated actual number exceeded the calculated number. This change gives the code official authority for the discretionary approval of lesser design occupant loads.

The code change allows the code official to be accommodating by recognizing the merit of the specific design while maintaining control of that design. For instance, the code official may want to create specific conditions for approval. While the exception allows for necessary design and approval flexibility, its use should be limited to very unique situations.

Although Table 1004.1.1 addresses a representative number of uses in the determination of a calculated occupant load, there are often uses that are not listed. New text gives the code official guidance as to the method of establishing the appropriate "occupant per unit of area" factor. The design occupant load determination should be based on the listed use that most nearly resembles the intended use, with the potential density of occupants, furnishings, fixtures, and equipment as the primary consideration.

Reorganization of the provisions has clarified the procedures by which design occupant loads are determined. This is particularly important because the accurate determination of design occupant loads is fundamental to the design of any means-of-egress system.

Table 1004.1.1

Occupant Load Determination for Day Care Uses

CHANGE TYPE. Addition

CHANGE SUMMARY. A new entry to Table 1004.1.1 provides for a density factor of 35 square feet per occupant for calculating the design occupant-load of a day care use, with the factor to be applied to the net floor area of the space. **(E7-04/05; E10-04/05)**

2006 CODE:

TABLE 1004.1.~~2~~1 Maximum Floor Area Allowances per Occupant

~~Occupancy~~Function of Space	Floor Area in Sq. Ft. per Occupant
Day care	35 net

(portions of table not shown did not change)

CHANGE SIGNIFICANCE. Previously, there was no guidance in the code for the determination of the design occupant load in a day care facility. Where such uses occurred, it was difficult to determine an appropriate factor from Table 1004.1.1 because day care was not specifically addressed, and none of the listed functions had similar density charac-

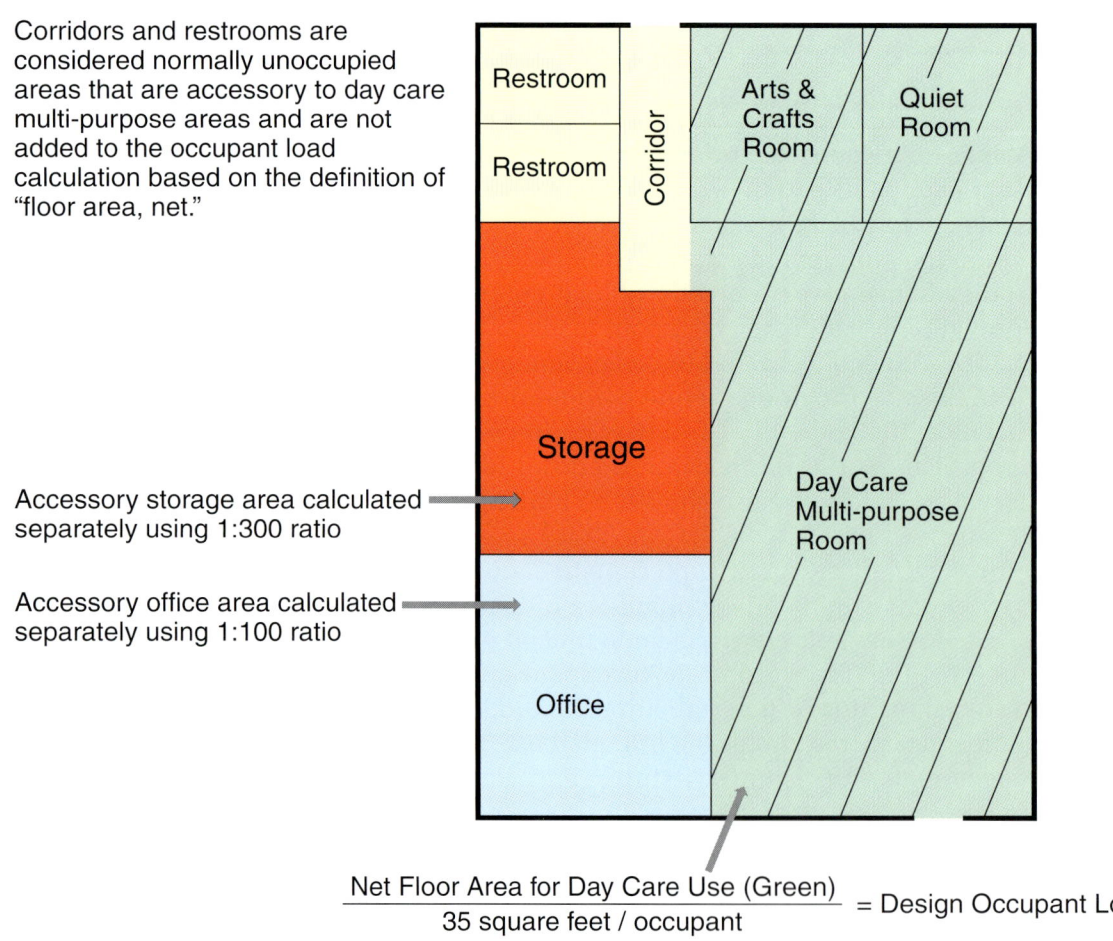

Corridors and restrooms are considered normally unoccupied areas that are accessory to day care multi-purpose areas and are not added to the occupant load calculation based on the definition of "floor area, net."

Accessory storage area calculated separately using 1:300 ratio

Accessory office area calculated separately using 1:100 ratio

$$\frac{\text{Net Floor Area for Day Care Use (Green)}}{35 \text{ square feet / occupant}} = \text{Design Occupant Load}$$

Occupant Load Determination for Day Care Facilities

teristics. The factor of 35 square feet per occupant is appropriate to address the anticipated number of children, furnishings, and equipment typically found in day care facilities.

The "net floor area" of the day care facility is the value to be used in applying the 35 square foot per occupant factor, which is limited to only the actual occupied area of the day care use. The floor areas of normally unoccupied spaces, such as toilet rooms, closets, and corridors, are not required to have a calculated occupant load in day care occupancies based on the definition of "floor area, net."

1004.2

Maximum Occupant Load Permitted

CHANGE TYPE. Modification

CHANGE SUMMARY. Where an occupant load is intended to exceed that established by Table 1004.1.1, the maximum occupant load shall be limited to one person per 7 square feet, resulting in a reduction in the total number of occupants that might be permitted in an assembly occupancy, which was previously based on a value of one person per 5 square fee. **(E9-03/04; E12-04/05)**

2006 CODE: **1004.2 Increased Occupant Load.** The occupant load permitted in any building, or portion thereof, is permitted to be increased from that number established for the occupancies in Table 1004.1.~~2~~ .1 provided that all other requirements of the code are also met based on such modified number and the occupant load shall not exceed one occupant per ~~5~~ 7 square feet (~~0.47~~ 0.65 m^2) of occupiable floor space. Where required by the building official, an approved aisle, seating, or fixed-equipment diagram substantiating any increase in occupant load shall be submitted. Where required by the building official, such diagram shall be posted.

Example:

Given: A casino floor having an area of 2695 square feet.

Determine: The maximum occupant load permitted for the casino floor.

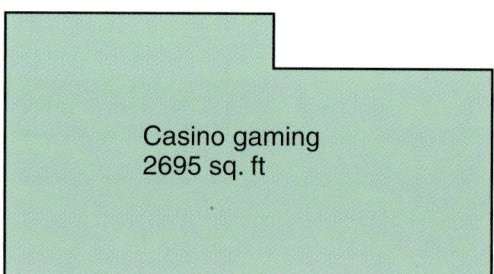

Solution:

$$\frac{2695 \text{ sq. ft}}{11 \text{ sq. ft. per occupant}} = 245 \quad \text{(Basic calculated design occupant load per Table 1004.1.1.)}$$

$$\frac{2695 \text{ sq. ft.}}{7 \text{ sq. ft. per occupant}} = 385 \quad \text{(Maximum occupant load that cannot be exceeded even where all code requirements applicable to an increased occupant load are satisfied. The previous maximum occupant load, based on 5 sq. ft. per person, would have been 539 occupants.)}$$

Maximum Occupant Load Calculation

CHANGE SIGNIFICANCE. The design occupant load is typically determined using values listed in Table 1004.1.1 based on the function and floor area of the space under consideration. The occupant load derived in this manner is typically used as the basis for design of the means of egress and in determining other code requirements that relate to occupant load.

The code permits the occupant load to be increased, but only where the means of egress is enhanced to accommodate the increased number of occupants and all other applicable requirements of the code, based on the increased occupant load, are satisfied.

Previously, the occupant density could be increased up to a maximum of 1 occupant per 5 square feet. This code change reduces the permitted occupant density to a maximum of 1 occupant per 7 square feet. The change was deemed appropriate to improve crowd movement in emergencies and addressed concerns about the ability of occupants in the center of a large, densely packed assembly area to safely access perimeter exits in a timely manner. The reduced occupant density also reduces the risk of jamming the means of egress in a mass evacuation.

1004.7

Occupant Load Determination for Fixed Seating

CHANGE TYPE.   Clarification

CHANGE SUMMARY.   Additional provisions clarify that where occupiable areas are located in conjunction with fixed seating, such areas must be included in the determination of the occupant load. **(E12-03/04)**

2006 CODE: 1004.7 Fixed Seating.   For areas having fixed seats and aisles, the occupant load shall be determined by the number of fixed seats installed therein. <u>The occupant load for areas in which fixed seating is not installed, such as waiting spaces and wheelchair spaces, shall be determined in accordance with Section 1004.1.1 and added to the number of fixed seats.</u>

For areas having fixed seating without dividing arms, the occupant load shall not be less than the number of seats based on one person for each 18 inches (457 mm) of seating length.

The occupant load of seating booths shall be based on one person for each 24 inches (610 mm) of booth seat length measured at the backrest of the seating booth.

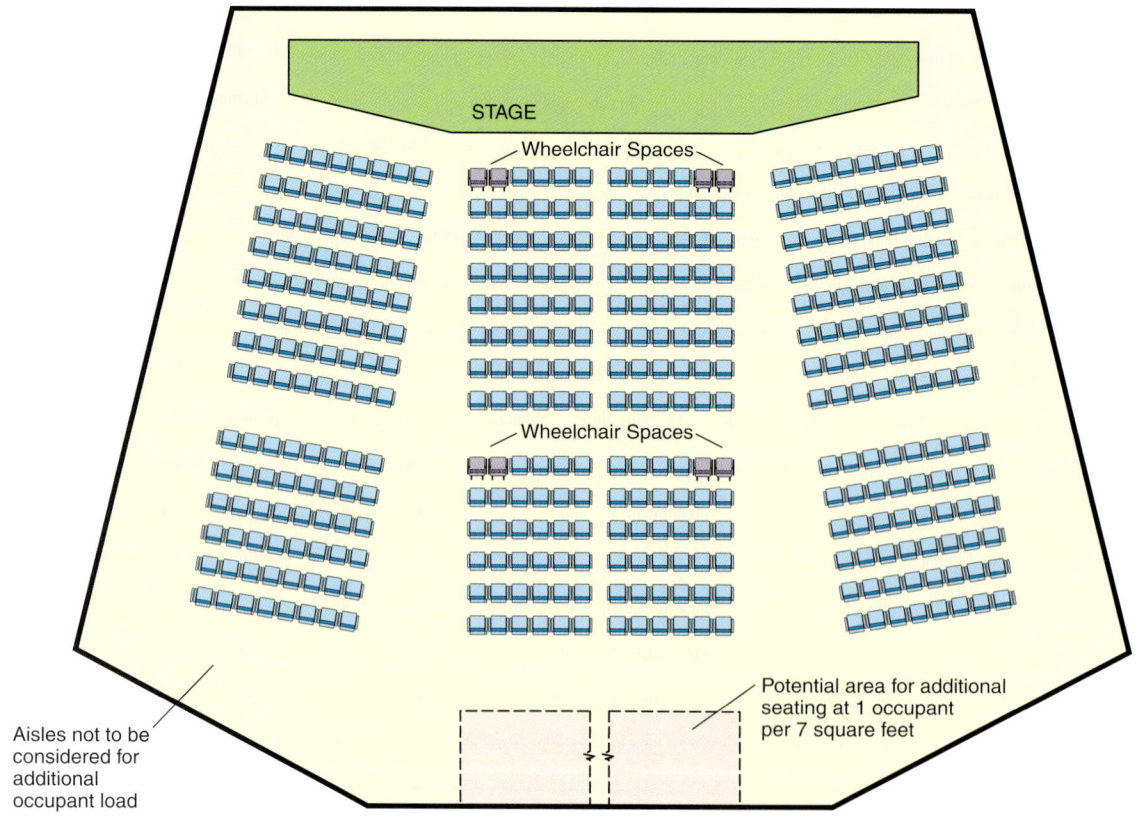

Occupant Load Determination for Fixed Seating

CHANGE SIGNIFICANCE. Under varying circumstances, fixed-seating assembly spaces may also contain other areas capable of being occupied. Such areas could include wheelchair spaces, waiting areas, and/or standing room. The new code text clarifies that the occupant load of any such areas must be added to that established for the fixed seating in the calculation of the occupant load. The inclusion of these additional occupiable areas will provide for a more accurate determination of the potential number of persons who could occupy the space.

1007.1

Platform Lifts as Accessible Means of Egress

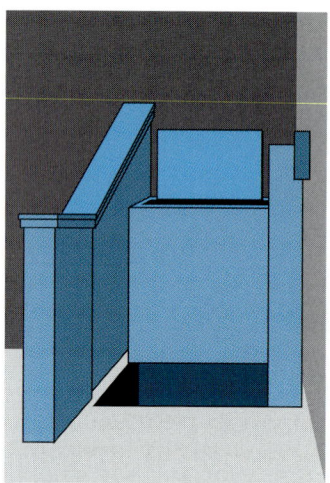

Platform Lifts as Part of an Accessible Means of Egress

CHANGE TYPE. Addition

CHANGE SUMMARY. A new exception permits a platform lift to be considered the single accessible means-of-egress component required from an accessible mezzanine level where the lift is permitted as a part of an accessible route. **(E16-03/04)**

2006 CODE: 1007.1 Accessible Means of Egress Required. Accessible means of egress shall comply with this section. Accessible spaces shall be provided with not less than one accessible means of egress. Where more than one means of egress is required by Section 1015.1 or 1019.1 from any accessible space, each accessible portion of the space shall be served by not less than two accessible means of egress.

Exceptions:
2. One accessible means of egress is required from an accessible mezzanine level in accordance with Section 1007.3, ~~or~~ 1007.4, <u>or 1007.5</u>.

1. and 3. (no change to text)

CHANGE SIGNIFICANCE. Platform lifts are permitted to be a portion of the accessible route under nine of the conditions set forth in Section 1109.7, several of which could address access to a small mezzanine level. The code change recognizes that if a platform lift is acceptable as an accessible route into a space, it should also be acceptable as an accessible means-of-egress element.

The general provisions require a minimum of two accessible means of egress from any accessible space required to have two or more means of egress. Exception 2 permits a single accessible means of egress from a mezzanine, regardless of the number of required means of egress. Previously, only a complying stairway or elevator could be utilized as the single accessible means of egress. The new reference to Section 1007.5 allows for the use of a platform lift under similar conditions. The allowance for use of a platform lift as a single accessible means of egress is strictly limited to mezzanine applications, and, more specifically, only where platform lift access is permitted by Section 1109.7.

In a related code change, Section 1007.5 prohibits the use of a platform lift as part of an accessible means of egress where the lift is used along the accessible route due to existing exterior site constraints. Due to the slow operation of a platform lift, it is considered inappropriate for egress purposes where there is the potential for a large number of lift users.

Section 1007.5.1 has been added to address the openness required of a platform lift. Lifts that are located on an accessible means of egress are no longer permitted to be installed within a full enclosure. It is important that the platform lift be open to allow for visual and audible communication with the user of the lift.

CHANGE TYPE.   Modification

CHANGE SUMMARY.   Where an enclosed exit stairway or an elevator is to be considered as a portion of an accessible means of egress, the requirement for providing an area of refuge or a horizontal exit as access to such a stairway or elevator is now mandated in fully sprinklered buildings as well as nonsprinklered buildings. **(E22-03/04; E23-04/05; E24-04/05; E28-04/05)**

2006 CODE: 1007.3 ~~Enclosed~~ **Exit Stairways.** ~~An enclosed exit stairway,~~ In order to be considered part of an accessible means of egress, an exit stairway shall have a clear width of 48 inches (1219 mm) minimum between handrails and shall either incorporate an area of refuge within an enlarged floor-level landing or shall be accessed from either an area of refuge complying with Section 1007.6 or a horizontal exit.

1007.3, 1007.4, 1007.6.2 continues

1007.3, 1007.4, 1007.6.2

Required Areas of Refuge

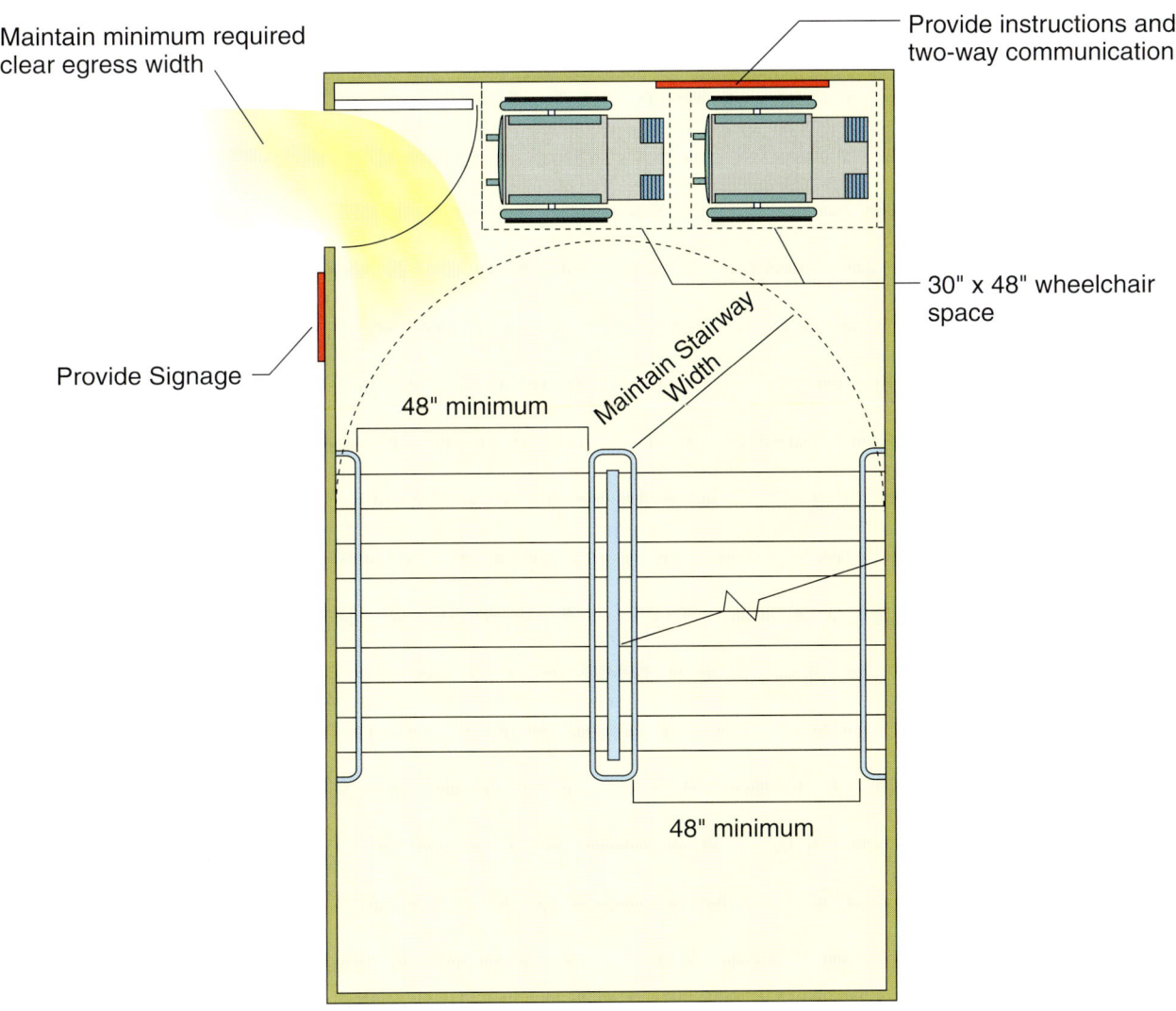

Area of Refuge in an Enclosed Stairway

Maintain minimum required clear egress width

Provide instructions and two-way communication

Provide Signage

30" x 48" wheelchair space

48" minimum

Maintain Stairway Width

48" minimum

1007.3, 1007.4, 1007.6.2 continued

Exceptions:

1. ~~Open~~ Unenclosed exit stairways as permitted by Section 10~~19~~20.1 are permitted to be considered part of an accessible means of egress.

2. The area of refuge is not required at ~~open~~ unenclosed exit stairways ~~that are~~ as permitted by Section 10~~19~~20.1 in buildings or facilities that are equipped throughout with an automatic sprinkler system in accordance with Section 903.3.1.1.

3. The clear width of 48 inches (1219 mm) between handrails ~~and the area of refuge~~ is not required at exit stairways in buildings or facilities equipped throughout with an automatic sprinkler system installed in accordance with Section 903.3.1.1 or 903.3.1.2.

4. **and 5.** (no significant changes to text)

1007.4 Elevators. ~~An elevator~~ In order to be considered part of an accessible means of egress, an elevator shall comply with the emergency operation and signaling device requirements of Section 2.27 of ASME A17.1. Standby power shall be provided in accordance with Sections 2702 and 3003. The elevator shall be accessed from either an area of refuge complying with Section 1007.6 or a horizontal exit.

Exceptions:

~~1.~~ Elevators are not required to be accessed from an area of refuge or horizontal exit in open parking garages.

~~2.~~ ~~Elevators are not required to be accessed from an area of refuge or horizontal exit in buildings and facilities equipped throughout with an automatic sprinkler system installed in accordance with Section 903.3.1.1 or 903.3.1.2.~~

1007.6.2 Separation. Each area of refuge shall be separated from the remainder of the story by a smoke barrier complying with Section 709 or a horizontal exit complying with Section 1021. Each area of refuge shall be designed to minimize the intrusion of smoke.

Exceptions:

~~1.~~ Areas of refuge located within a ~~stairway~~ vertical exit enclosure.

~~2.~~ ~~Areas of refuge where the area of refuge and areas served by the area of refuge are equipped throughout with an automatic sprinkler system installed in accordance with Section 903.3.1.1 or 903.3.1.2.~~

CHANGE SIGNIFICANCE. Enclosed exit stairways and elevators are components often utilized to provide for an accessible means of egress. As a condition of their use as accessible means of egress components, they are required to be provided with an area of refuge or horizontal exit that will provide for a level of protection from fire and smoke. In the previous code, buildings that were provided with an automatic sprinkler system throughout did not require horizontal exits or areas of refuge at enclosed exit stairways or elevators. The sprinkler exemption has been deleted, and now the requirements are consistent for both nonsprinklered buildings and fully sprinklered buildings.

The previous exception for fully sprinklered buildings was based on the recognition of an increased level of safety and evacuation time that is afforded in a sprinklered building. The expectation was that a supervised system would reduce the threat of fire by reliably controlling and confining the fire to the immediate area or origin. The revised provision reflects a concern that the sprinkler trade-off may not be appropriate for omitting the area of refuge or horizontal exit utilized in an accessible means of egress for mobility-impaired occupants.

1008.1.1

Minimum Door Width in Group R-1 Occupancies

CHANGE TYPE. Modification

CHANGE SUMMARY. Interior doors within dwelling units and sleeping units of Group R-1 occupancies are no longer exempt from the general provisions regulating door width, resulting in a minimum door-opening width of 32 inches. **(E2-03/04; E33-04/05; E34-04/05)**

2006 CODE: **1008.1.1 Size of Doors.** The minimum width of each door opening shall be sufficient for the occupant load thereof and shall provide a clear width of not less than 32 inches (813 mm). Clear openings of doorways with swinging doors shall be measured between the face of the door and the stop, with the door open 90 degrees (1.57 rad). Where this section requires a minimum clear width of 32 inches (813 mm) and a door opening includes two door leaves without

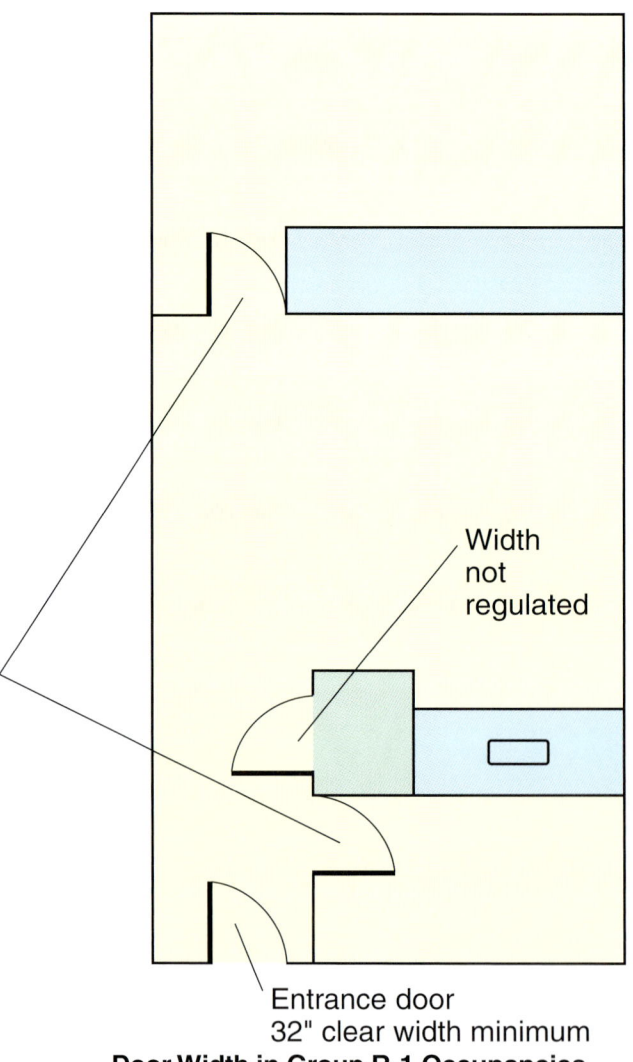

Group R-1
Dwelling Unit or
Sleeping Unit*

*Not required to be an
Accessible, Type A or
Type B unit.

Width
not
regulated

Interior
passage doors
32" clear width
minimum

Entrance door
32" clear width minimum

Door Width in Group R-1 Occupancies

a mullion, one leaf shall provide a clear opening width of 32 inches (813 mm). The maximum width of a swinging door shall be 48 inches (1219 mm) nominal. Means of egress doors in an occupancy in Group 1-2 used for the movement of beds shall provide a clear width not less than 41½ inches (1054 mm). The height of doors shall not be less than 80 inches (2032 mm).

Exceptions:

1. through 6. (no change to text)

7. ~~Interior egress doors within a dwelling unit or sleeping unit which is not required to be adaptable or accessible~~ In other than Group R-1 occupancies, the minimum widths shall not apply to interior egress doors within a dwelling unit or sleeping unit that is not required to be an Accessible unit, Type A unit, or Type B unit.

8. Door openings required to be accessible within Type B ~~dwelling~~ units shall have a minimum clear width of 31.75 inches (806 mm).

CHANGE SIGNIFICANCE. Door openings are generally required to provide a minimum of 32 inches of clear width. Previously, Exception 7 exempted interior egress doors within dwelling units or sleeping units if such units were not considered as Accessible, Type A or Type B. As a result of this code change, the exception is no longer applicable to units that are classified as Group R-1 occupancies. Therefore, all interior egress doorways within a Group R-1 dwelling unit or sleeping unit must provide at least 32 inches in clear width.

The requirement for all doorways within a Group R-1 unit to be sized to provide access to the physically disabled is applicable to both entrance doors and passage doors. Because of the social interaction and visitation that often occur in lodging facilities, a door opening sized for accessibility is deemed necessary to allow people with disabilities to visit a friend's, colleague's, or relative's unit that may not necessarily be one of the required accessible units. In addition, wider doors provide an added benefit to all persons handling luggage and bulky items.

A clarification to Exception 8 indicates that the permitted reduction in door-opening width to 31¾ inches previously allowed for Type B dwelling units is also applicable to Type B sleeping units.

1008.1.2

Door Swing in Sleeping Units

CHANGE TYPE. Addition

CHANGE SUMMARY. The requirement that egress doors be side-hinged swinging doors is no longer applicable to patient rooms in health care suites, as well as bathroom doors of hotel guest rooms and other Group R-1 sleeping units. **(E22-03/04; E36-04/05)**

2006 CODE: 1008.1.2 Door Swing. Egress doors shall be side-hinged swinging.

> **Exceptions:**
> 1. and 2. (no change to text)
> 3. Critical or intensive care patient rooms within suites of health care facilities.
> 3. 4. Doors within or serving a single dwelling unit in Groups R-2 and R-3 as applicable in Section 101.2.
> 4. 5. In other than Group H occupancies, revolving doors complying with Section 1008.1.3.1.
> 5. 6. In other than Group H occupancies, horizontal sliding doors complying with Section 1008.1.3.3 in a means of egress.
> 6. 7. Power-operated doors in accordance with Section 1008.1.3.1.
> 8. Doors serving a bathroom within an individual sleeping unit in Group R-1.

CHANGE SIGNIFICANCE. In critical care or intensive care patient rooms, the use of egress doors other than side-hinged swinging doors had previously been limited primarily to power-operated doors permitted under Exception 7. This change responds to a desire to utilize sliding glass doors in such patient rooms to allow for visual observation and the efficient movement of equipment, rather than swinging doors. As patients within these areas of health care suites would not

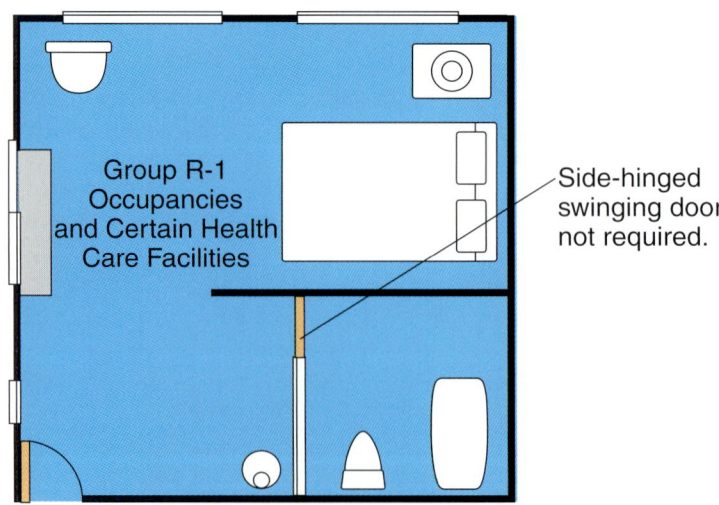

Permissible Use of Non-Side Hinged Doors

initially be relocated in an emergency because of the defend-in-place philosophy, it was determined that requiring power-operated doors in all of the patient rooms of such suites was overly restrictive. Therefore, a new exception mandates no requirement for swinging doors in such a controlled and limited use.

It is often beneficial to utilize sliding pocket doors to provide access to bathrooms in hotel guest rooms as well, particularly in those required to be accessible. Conflicts often occur between door swing and the required clearances for plumbing fixtures. This change also allows the use of sliding doors, as well as any other type of door, for this specific application in Group R-1.

Use of a door style other than the side-hinged, swinging type is deemed acceptable because of the limited occupant load served by the door, the limited travel distance to a corridor or similar exitway, and a similar exception applicable to dwelling units in Group R-2 and R-3 occupancies. The exception applies only to the method of door movement and does not modify any other applicable requirements, such as doorway width and door operation.

1008.1.6

Thresholds at Residential Exterior Doors

CHANGE TYPE.   Modification

CHANGE SUMMARY.   In order to provide for a threshold height of up to 7¾ inches at an exterior door that is not a portion of the means of egress in Group R-2 occupancies, the door can no longer swing over an exterior landing or step. **(E25-03/04; E26-03/04)**

2006 CODE: 1008.1.6 Thresholds.   Thresholds at doorways shall not exceed 0.75 inch (19.1 mm) in height for sliding doors serving dwelling units or 0.5 inch (12.7 mm) for other doors. Raised thresholds and floor level changes greater than 0.25 inch (6.4 mm) at doorways shall be beveled with a slope not greater than one unit vertical in two units horizontal (50% slope).

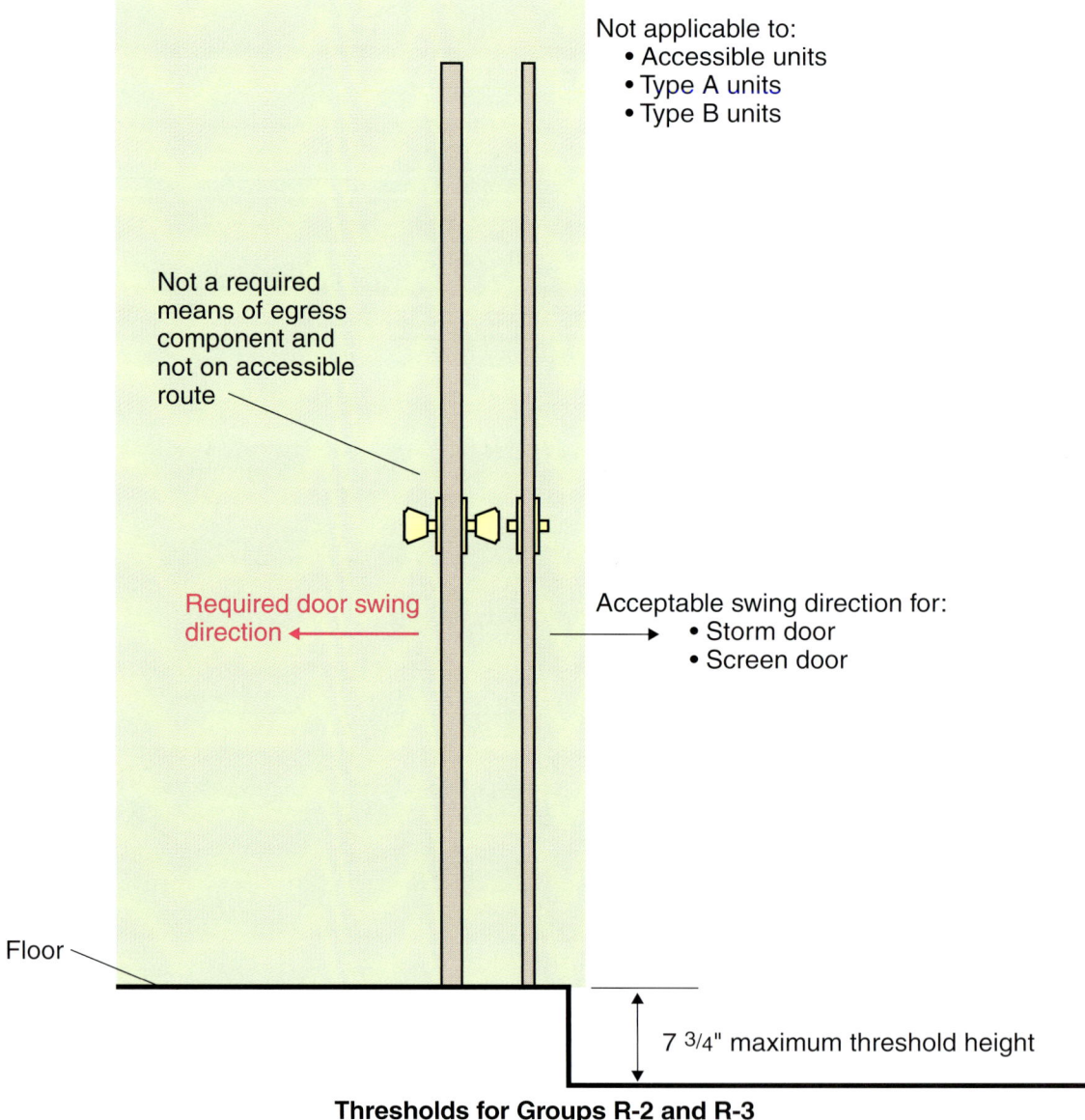

Not applicable to:
- Accessible units
- Type A units
- Type B units

Not a required means of egress component and not on accessible route

Required door swing direction ←

Acceptable swing direction for:
- Storm door
- Screen door

Floor

7 3/4" maximum threshold height

Thresholds for Groups R-2 and R-3

Exception: The threshold height shall be limited to 7.75 inches (197 mm) where the occupancy is Group R-2 or R-3 ~~as applicable in Section 101.2,~~; the door is an exterior door that is not a component of the required means of egress~~,~~; <u>the door, other than an exterior storm or screen door, does not swing over the landing or step;</u> and the doorway is not on an accessible route <u>as required by Chapter 11 and the door is not part of an Accessible unit, Type A unit, or Type B unit.</u>

CHANGE SIGNIFICANCE. Although thresholds are typically limited to a maximum height of ½ inch, an increased height has been permitted for exterior doors in Groups R-2 and R-3 that are not components of the required means of egress. Previously, the direction of door swing was not a consideration for application of the increased height. The modified code text now mandates that the $7\frac{3}{4}$-inch allowance is permitted only where the door does not swing over the exterior landing or step. There is a considerably greater hazard involved where travel occurs through a door that swings over a lowered walking surface. This limitation is not applicable to exterior storm doors or screen doors. The change also provides for coordination with Exception 3 of Section 1008.1.4, addressing Group R-3 occupancies.

1008.1.8.7

Remote Unlocking of Stairway Doors

CHANGE TYPE. Modification

CHANGE SUMMARY. An electronic override is now required for the unlocking of stairway doors from the stairway side in stairways serving four or fewer stories. **(E29-03/04)**

2006 CODE: 1008.1.8.7 Stairway Doors. Interior stairway means of egress doors shall be openable from both sides without the use of a key or special knowledge or effort.

Exceptions:

1. Stairway discharge doors shall be openable from the egress side and shall only be locked from the opposite side.
2. This section shall not apply to doors arranged in accordance with Section 403.12.
3. In stairways serving not more than four stories, doors are permitted to be locked from the side opposite the egress side, provided they are openable from the egress side <u>and capable of being unlocked simultaneously without unlatching upon a signal from the fire command center, if present, or a signal by emergency personnel from a single location inside the main entrance to the building.</u>

Electronic override is required to allow remote unlocking of many stairway door locks.

CHANGE SIGNIFICANCE. Where stairways serve four or fewer stories, the code has previously permitted the locking of the stairway doors from the stairway side. There were no limitations placed on the use of the exception. This has provided a high degree of security for building owners and occupants, but it creates conditions detrimental to firefighters and emergency responders. The code change maintains the allowance to lock the stairway doors; however, a means of simultaneously unlocking all of the doors by emergency personnel must now be provided.

The provision further requires that the stairway doors be unlocked without unlatching. Because the stairway doors will typically be fire-door assemblies, latching is necessary to maintain the integrity of the fire-resistive separation between the exit enclosure and the remainder of the building.

The remote unlocking signal must be initiated from the fire command center, in those limited number of buildings where such a facility is provided. Typically, the single point of signal initiation will occur at an approved location inside the building's main entrance.

1008.1.9

Panic and Fire Exit Hardware

CHANGE TYPE. Modification

CHANGE SUMMARY. The threshold requirement for panic hardware in Group A and E occupancies has been reduced from 100 occupants to 50. In addition, a new exception clarifies that panic hardware is not mandated in Group A occupancies on egress doors where key-operated locking devices are installed in accordance with Section 1008.1.8.3, Item 2. Another new provision requires the use of panic hardware on egress doors from specified electrical rooms. **(E31-03/04; E33-03/04; E34-03/04; E1-04/05)**

2006 CODE: 1008.1.9 Panic and Fire Exit Hardware. Where panic and fire exit hardware is installed, it shall comply with the following:

1. The actuating portion of the releasing device shall extend at least one-half of the door leaf width.

2. ~~A~~ The maximum unlatching force ~~of~~ shall not exceed 15 pounds (67 N).

Each door in a means of egress from ~~an occupancy of~~ a Group A or E occupancy having an occupant load of ~~100~~ 50 or more and any ~~occupancy of~~ Group H ~~1, H 2, H 3 or H 5~~ occupancy shall not be provided with a latch or lock unless it is panic hardware or fire exit hardware.

Exception: A main exit of a Group A occupancy in compliance with Section 1008.1.8.3, Item 2.

Electrical rooms with equipment rated 1200 amperes or more and over 6 feet (1829 mm) wide that contain overcurrent devices, switching devices, or control devices, with exit access doors must be equipped with panic hardware and doors must swing in the direction of egress.

Panic hardware required as locking/latching device where egress door serves:
- 50 or more occupants in Groups A and E
- Any occupant load in Group H
- Certain electrical rooms

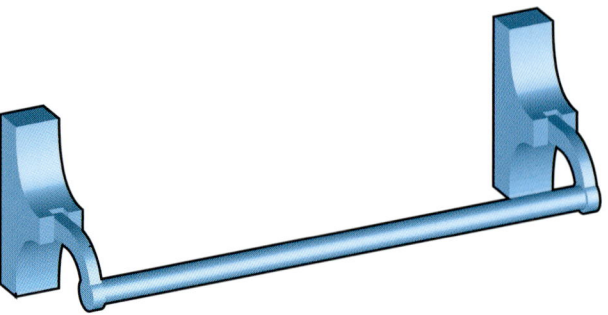

Panic Hardware

If balanced doors are used and panic hardware is required, the panic hardware shall be the push-pad type and the pad shall not extend more than one-half the width of the door measured from the latch side.

CHANGE SIGNIFICANCE.   The change of the design occupant load at which panic hardware is required in Group A and E occupancies correlates with the present provisions requiring a second means of egress. It also enhances the level of life safety for crowds as small as 50 by providing door hardware that will prevent a door from being jammed as a result of pressure against a door during emergency egress situations. The addition of the exception makes it clear that the provisions for key-locking hardware at the main exit in Group A occupancies take precedence over those of Section 1008.1.9 for panic hardware.

Traditionally, means-of-egress doors in Group H occupancies containing physical hazards (Groups H-1, H-2, and H-3) have been required to have panic hardware where a lock or latch is utilized. The code now also mandates such hardware for doors in Group H-4 occupancies, where health hazards are the primary concern. Because the effects of acute health hazard materials, such as toxic and highly toxic materials, may diminish the physical and mental capacities of the occupants, it is important that egress doors be easily operable. Arguably, this new requirement is overly restrictive for solid and liquid materials, particularly corrosive materials, but industry did not offer significant opposition to the proposal, which might have addressed these issues.

The new provisions addressing panic hardware on electrical-room doors are intended to provided consistency with similar provisions in the National Electrical Code. The requirement is applicable only where multiple conditions are present. Because the type of room regulated by this requirement creates a potentially hazardous environment in the event of an electrical accident, the more-immediate egress provided by panic hardware is considered to be desirable.

1009.3.1, 1009.7

Curved Stairways

CHANGE TYPE. Modification

CHANGE SUMMARY. "Circular stairways" have been renamed "curved stairways" to allow for the utilization of other types of stair configurations. The minimum required radius of a curved stairway is now based on the required stairway width rather than the actual width. **(E40-03/04; E50-04/05)**

2006 CODE: ~~1009.8~~ 1009.3.1 ~~Winders~~ Treads. Winder treads are not permitted in means of egress stairways except within a dwelling unit.

> **Exceptions:** (from previous Section 1009.3)
> 1. ~~Circular~~ Curved stairways in accordance with Section 1009.7.
> 2. Spiral stairways in accordance with Section 1009.8.

1009.7 ~~Circular~~ Curved Stairways. ~~Circular~~ Curved stairways with winder treads shall have ~~a minimum~~ treads ~~depth~~ and ~~a maximum~~ risers ~~height~~ in accordance with Section 1009.3 and the ~~smaller~~ smallest radius shall not be less than twice the required width of the stairway. ~~The minimum tread depth measured 12 inches (305 mm) from the narrower end of the tread shall not be less than 11 inches (279 mm). The minimum tread depth at the narrow end shall not be less than 10 inches (254 mm).~~

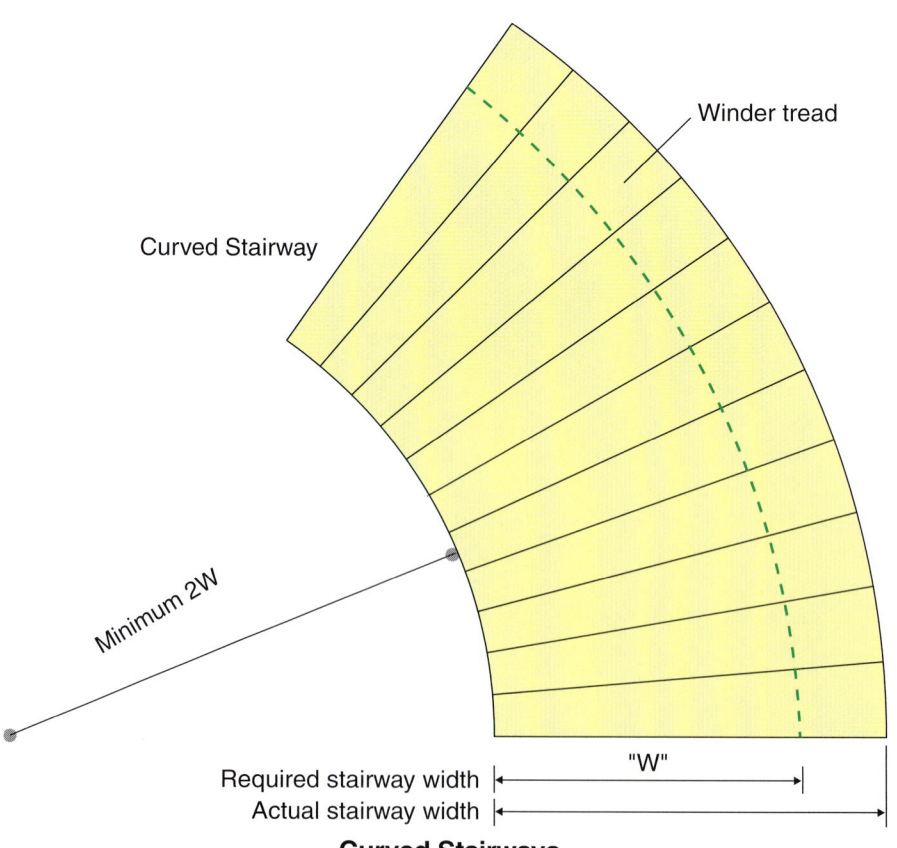

Curved Stairways

Exception: <u>The radius restriction shall not apply to curved</u> <u>stairways</u> for occupancies in Group R-3, and within individual dwelling units in occupancies in Group R-2.~~, both as applicable in Section 101.2.~~

CHANGE SIGNIFICANCE. "Circular stairs," now described by the code as "curved stairways," are designed by arraying the winder treads in a circular plan like the spokes of a wheel. Previously, the minimum required inside radius for a curved (circular) stairway was determined by doubling the *actual* width of the stairway. The new method set forth in the code change mandates that the smallest radius must be at least twice the *required* width of the stairway, often resulting in a tighter permissible radius. The minimum-radius limitation basically restricts the angularity of the uniform winder treads in circular stairways and limits the turn that users experience as they travel on the stairway. The revised method of measurement provides for a more consistent application of the stairway criteria. The minimum required radius would now typically be established at 88 inches, because most curved stairways have a minimum required width of 44 inches. As the occupant load served by the stair increases to the point where the required width of the stair increases, the result will be an increased inside radius, which further reduces the angularity.

By replacing the term "circular" with "curved," the code change reflects recognition that other types of curved stairways can be safely designed for egress situations with the radius restriction mandated by the code. Elliptical curves and combinations of radii can be effectively used to design curved stairs that are safe because of the limit on the smallest radius, creating a comfortably turning stair regardless of the type of curve or combination of curves used by the designer.

Further modifications to the previous code text are an acknowledgment that all of the tread and riser parameters of Section 1009.3 apply, not just those addressing tread depth and riser height. Recognition that circular stairways contain winder treads is established, whereas the exception for individual dwelling units in Groups R-2 and R-3 is limited to the radius-restriction provision.

1009.5.2, 1010.7.2, 1014.5

Weather Protection of Exterior Egress Components

CHANGE TYPE. Deletion

CHANGE SUMMARY. The requirement that outdoor stairways, ramps, and egress balconies be protected in a manner to minimize or prevent the accumulation of ice and snow has been deleted. **(E46-03/04)**

2006 CODE: **1009.5.2 Outdoor Conditions.** Outdoor stairways and outdoor approaches to stairways shall be designed so that water will not accumulate on walking surfaces. ~~In other than occupancies in Group R-3, and occupancies in Group U that are accessory to an occupancy in Group R-3, treads, platforms and landings that are part of exterior stairways in climates subject to snow or ice shall be protected to prevent the accumulation of same.~~

1010.7.2 Outdoor Conditions. Outdoor ramps and outdoor approaches to ramps shall be designed so that water will not accumulate on walking surfaces. ~~In other than occupancies in Group R-3, and occupancies in Group U that are accessory to an occupancy in Group R-3, surfaces and landings that are part of exterior stairways in climates subject to snow or ice shall be protected to minimize the accumulation of same.~~

Exterior egress components, such as these stairways, no longer require protection to minimize or prevent the accumulation of snow and ice.

1014.5 Egress Balconies. Balconies used for egress purposes shall conform to the same requirements as corridors for width, headroom, dead ends, and projections. ~~Exterior balconies shall be designed to minimize accumulation of snow or ice that impedes the means of egress.~~

> **Exception:** ~~Exterior balconies and concourses in outdoor stadiums shall be exempt from the design requirement to protect against the accumulation of snow or ice.~~

CHANGE SIGNIFICANCE. Various code provisions continue to maintain the requirement that outdoor walking surfaces, such as exterior stairways, ramps, and egress balconies, be provided with a means for water drainage. This is typically accomplished by providing a slight slope to the surface such that water will not accumulate. However, the previous requirements for providing a means to minimize or prevent the accumulation of snow or ice have been removed.

In many climates, only completely enclosing the exterior egress element can prevent ice or snow accumulation on a walking surface. Even the installation of coverings over outdoor stairways, ramps, egress balconies, and their approaches cannot totally prevent accumulation. In addition, provisions that required "minimizing" ice or snow accumulation were considered too subjective for use as a basis for a code requirement with such a broad impact.

1009.5.3

Enclosed Usable Space under Stairways

CHANGE TYPE. Addition

CHANGE SUMMARY. Where enclosed usable space is created under a stairway that is located within a dwelling unit classified as a Group R-2 or R-3 occupancy, the enclosure shall be protected with minimum ½-inch gypsum board on the enclosed side. **(E96-03/04)**

2006 CODE: ~~**1019.1.5**~~ **1009.5.3 Enclosures Under Stairways.** The walls and soffits within enclosed usable spaces under enclosed and unenclosed stairways shall be protected by 1-hour fire-resistance-rated construction, or the fire-resistance rating of the stairway enclosure, whichever is greater. Access to the enclosed usable space shall not be directly from within the stair enclosure.

> **Exception:** Spaces under stairways serving and contained within a single residential dwelling unit in Group R-2 or R-3 ~~as applicable in Section 101.2~~ <u>shall be permitted to be protected on the enclosed side with 0.5-inch (12.7 mm) gypsum board.</u>

There shall be no enclosed usable space under exterior exit stairways unless the space is completely enclosed in 1-hour fire-resistance-

When storage areas beneath stairs in Group R-2 or Group R-3 dwelling units are enclosed, the enclosure must be constructed using minimum ½-inch gypsum board.

rated construction. The open space under exterior stairways shall not be used for any purpose.

CHANGE SIGNIFICANCE.   It is not uncommon for the space under a stairway within a dwelling unit to be enclosed for storage or similar purposes. If a fire within the enclosed space was to go undetected for some time, the stairway would potentially become unusable for egress. The code previously required no fire-resistive protection on the interior of such enclosed, usable space, but now, the enclosed side must now be protected with minimum $\frac{1}{2}$-inch gypsum board. It is assumed that the increased protection will provide more time for smoke to be detected before the stair is compromised by fire.

1009.10, 1010.8, 1012

Handrails for Stairways and Ramps

CHANGE TYPE.   Clarification

CHANGE SUMMARY.   The provisions regulating handrails have been moved from within the stairway requirements to their own section, because handrails are required for both stairways and ramps. **(E53-04/05; E54-04/05)**

2006 CODE: **1009.10 Handrails.**   Stairways shall have handrails on each side and shall comply with Section 1012. Where glass is used to provide the handrail, the handrail shall also comply with Section 2407 of the *International Building Code.*

> **Exceptions:**
> 1. through 5. (no changes to the current text)

1010.8 Handrails.   Ramps with a rise greater than 6 inches (152 mm) shall have handrails on both sides. Handrails shall comply ~~complying~~ with Section ~~1009.11~~ 1012.

SECTION 1012 HANDRAILS

1012.1 Where Required.   Handrails for stairways and ramps shall be adequate in strength and attachment in accordance with Section 1607.7. Handrails required for stairways by Section 1009.10 shall comply with Section 1012.2 through 1012.8. Handrails required for ramps by Section 1010.8 shall comply with Sections 1012.2 through 1012.7.

~~1009.11.1~~ **1012.2 Height.**   Handrail height, measured above stair tread nosings, or finish surface of ramp slope, shall be uniform, not less than 34 inches (864 mm) and not more than 38 inches (965 mm).

Requirements for handrails have been clarified.

~~1009.11.3~~ <u>1012.3</u> **Handrail Graspability.**   Handrails with a circular cross section shall have an outside diameter of at least 1.25 inches (32 mm) and not greater than 2 inches (51 mm) or shall provide equivalent graspability. If the handrail is not circular, it shall have a perimeter dimension of at least 4 inches (102 mm) and not greater than 6.25 inches (160 mm) with a maximum cross-section dimension of 2.25 inches (57 mm). Edges shall have a minimum radius of 0.01 inch (0.25 mm).

~~1009.11.4~~ <u>1012.4</u> **Continuity.**   Handrail-gripping surfaces shall be continuous, without interruption by newel posts or other obstructions.

Exceptions:
1. Handrails within dwelling units are permitted to be interrupted by a newel post at a stair <u>or rail</u> landing.
2. Within a dwelling unit, the use of a volute, turnout or staring easing is allowed on the lowest tread.
3. Handrail brackets or balusters attached to the bottom surface of the handrail that do not project horizontally beyond the sides of the handrail with 1.5 inches (38 mm) of the bottom of the handrail shall not be considered to be obstructions and provided further that for each 0.5 inch (13 mm) of additional handrail perimeter dimension above 4 inches (102 mm), the vertical clearance dimension of 1.5 inches (38 mm) shall be permitted to be reduced by 0.125 inch (3 mm).

~~1009.11.5~~ <u>1012.5</u> **Handrail Extensions.**   Handrails shall return to a wall, guard, or the walking surface or shall be continuous to the handrail of an adjacent stair flight <u>or ramp run. At stairways</u> where handrails are not continuous between flights, the handrails shall extend horizontally at least 12 inches (305 mm) beyond the top riser and continue to slope for the depth of one tread beyond the bottom riser. <u>At ramps where handrails are not continuous between runs, the handrail shall extend horizontally above the landing 12 inches (305 mm) minimum beyond the top and bottom ramps.</u>

Exceptions:
1. Handrails within a dwelling unit that is not required to be accessible need extend only from the top riser to the bottom riser.
2. Aisle handrails in Group A occupancies in accordance with Section 1025.13.

~~1009.11.6~~ <u>1012.6</u> **Clearance.**   Clear space between a handrail and a wall or other surface shall be a minimum of 1.5 inches (38 mm). A handrail and a wall or other surface adjacent to the handrail shall be free of any sharp or abrasive elements.

~~1109.11.6 Stairway~~ <u>1012.7</u> **Projections.**   <u>On ramps, the clear width between handrails shall be 36 inches (914 mm) minimum.</u> Projections into the required width <u>of stairways and ramps</u> at each handrail shall not exceed 4.5 inches (114 mm) at or below the handrail

1009.10, 1010.8, 1012 continues

1009.10, 1010.8, 1012 continued

height. Projections into the required width shall not be limited above the minimum headroom height required in Section 1009.2.

~~**1009.11.2**~~ **1012.8 Intermediate Handrails.** <u>Stairways shall have</u> intermediate handrails ~~are required~~ <u>located in such a manner</u> so that all portions of the stairway width required for egress capacity are within 30 inches (762 mm) of a handrail. On monumental stairs, handrails shall be located along the most direct path of egress travel.

CHANGE SIGNIFICANCE.   Handrails are mandated by the code for both stairways and ramps; however, the requirements for handrails were previously only within the stairway provisions. This created the potential for confusion and misapplication. In the new format, handrails are located in their own section, similar to that for guards. The relocation also allows for recognition that some handrail provisions for ramps are different from those for stairways, such as the requirement for intermediate handrails. Correlation with the ramp requirements of ICC A117.1 has also been accomplished.

1009.11.2, 1013.5, 1013.6

Protection at Roof-Hatch Openings

CHANGE TYPE. Addition

CHANGE SUMMARY. A complying guard is now required where a roof-hatch opening is located close to the roof edge. In addition, the minimum extent of any required rooftop guard is now established. **(E51-03/04; E67-03/04)**

2006 CODE: 1009.11.2 Protection at Roof Access Openings. Where the roof hatch opening providing the required access is located within 10 feet (3049 mm) of the roof edge, such roof access or roof edge shall be protected by guards installed in accordance with the provisions of Section 1013.

~~1012.5~~ **1013.5 Mechanical Equipment.** Guards shall be provided where appliances, equipment, fans, roof hatch openings, or other components that require service are located within 10 feet (3048 mm) of a roof edge or open side of a walking surface and such edge or open side is located more than 30 inches (762 mm) above the floor, roof, or grade below. The guard shall be constructed so as to prevent the passage of a 21-inch-diameter (533 mm) sphere. The guard shall extend not less than 30 inches (762 mm) beyond each end of such appliance, equipment, fan, or component.

1013.6 Roof Access. Guards shall be provided where the roof hatch opening is located within 10 feet (3048 mm) of a roof edge or open side of a walking surface and such edge or open side is located more than 30 inches (762 mm) above the floor, roof, or grade below. The guard shall be constructed so as to prevent the passage of a 21-inch-diameter (533 mm) sphere.

1009.11.2, 1013.5, 1013.6 continues

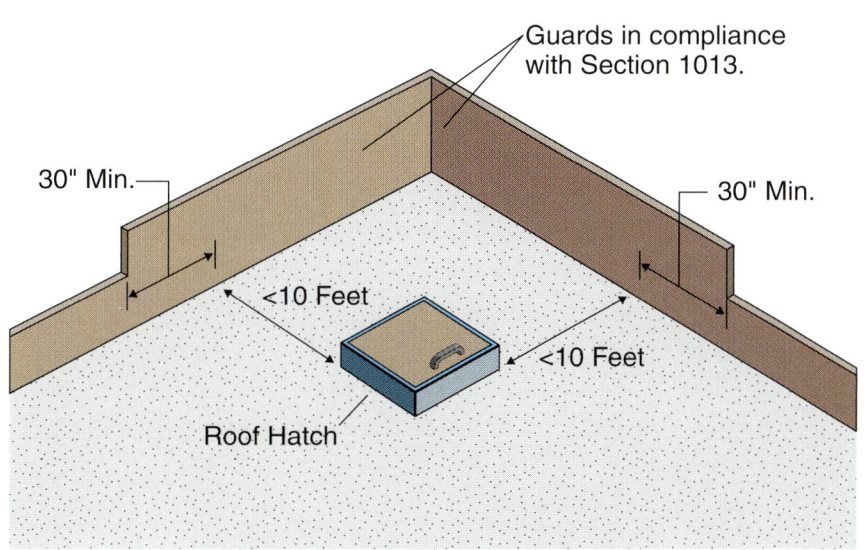

Protection at Roof-Hatch Openings

1009.11.2, 1013.5, 1013.6 continued

CHANGE SIGNIFICANCE. Roof hatches are permitted as a means of access to unoccupied roofs where such access is required in buildings four or more stories in height. In addition, roof-hatch openings are often provided for other purposes, including access to rooftop equipment. These new code provisions address the hazard created where the roof hatch is located very close to the roof edge. Although the code has long required guards adjacent to normally utilized elevated walking surfaces, it did not specifically address unoccupied limited-use roof areas. The new text provides a minimum level of safety for emergency responders, service personnel, inspectors, and others.

The provisions of Section 1009.11 mandate various methods for access to the roof of any building four or more stories in height. If the roof is not considered typically occupiable, the access may include a roof hatch. Where a roof hatch is utilized to satisfy the access requirements of Section 1009.11, guard protection is mandated where the hatch is located within 10 feet of any roof edge.

Section 1013.5 has historically required guards where appliances, equipment, fans, or other mechanical equipment are located adjacent to a roof edge. This provision primarily addressed the hazard created when service personnel are working on the rooftop equipment. The added text recognizes that a hatch opening located near the edge of the roof presents similar concerns. It is appropriate to protect persons accessing the roof by a hatch from falling off a roof as a result of a trip or misstep. At times, these roof accesses are used during inclement weather, emergency situations, or darkness. In addition, the area around roof-hatch openings is also often utilized as a staging area or work area. The code now requires guards to extend 30 inches beyond the edge of the rooftop component requiring the guard. The criteria for openings in any required guards, which prohibit the passage of a 21-inch-diameter sphere, are consistent with those for other elevated areas not accessible to the public.

Note that Sections 1013.5 and 1013.6 overlap because of an apparent failure to coordinate the sections when the code was being revised. The reference to roof hatch openings in Section 1013.5 was originally intended to apply to hatches used for equipment access, but the final code text lost this limitation. Without it, provisions in Section 1013.5 make those in Section 1013.6 unnecessary.

CHANGE TYPE. Modification

CHANGE SUMMARY. The minimum required length of a ramp landing has been reduced from 60 inches to 48 inches for any ramp that is not a part of an accessible route. **(E53-03/04)**

1010.6.3
Minimum Ramp Length

2006 CODE: 1010.6.3 Length. The landing length shall be 60 inches (1525 mm) minimum.

Exceptions:
1. Landings in nonaccessible Group R-2 and R-3 individual dwelling units, as applicable in Section 1001.1, are permitted to be 36 inches (914 mm) minimum.

2. Where the ramp is not a part of an accessible route, the length of the landing shall not be required to be more than 48 inches (1220 mm) in the direction of travel.

1010.6.3 continues

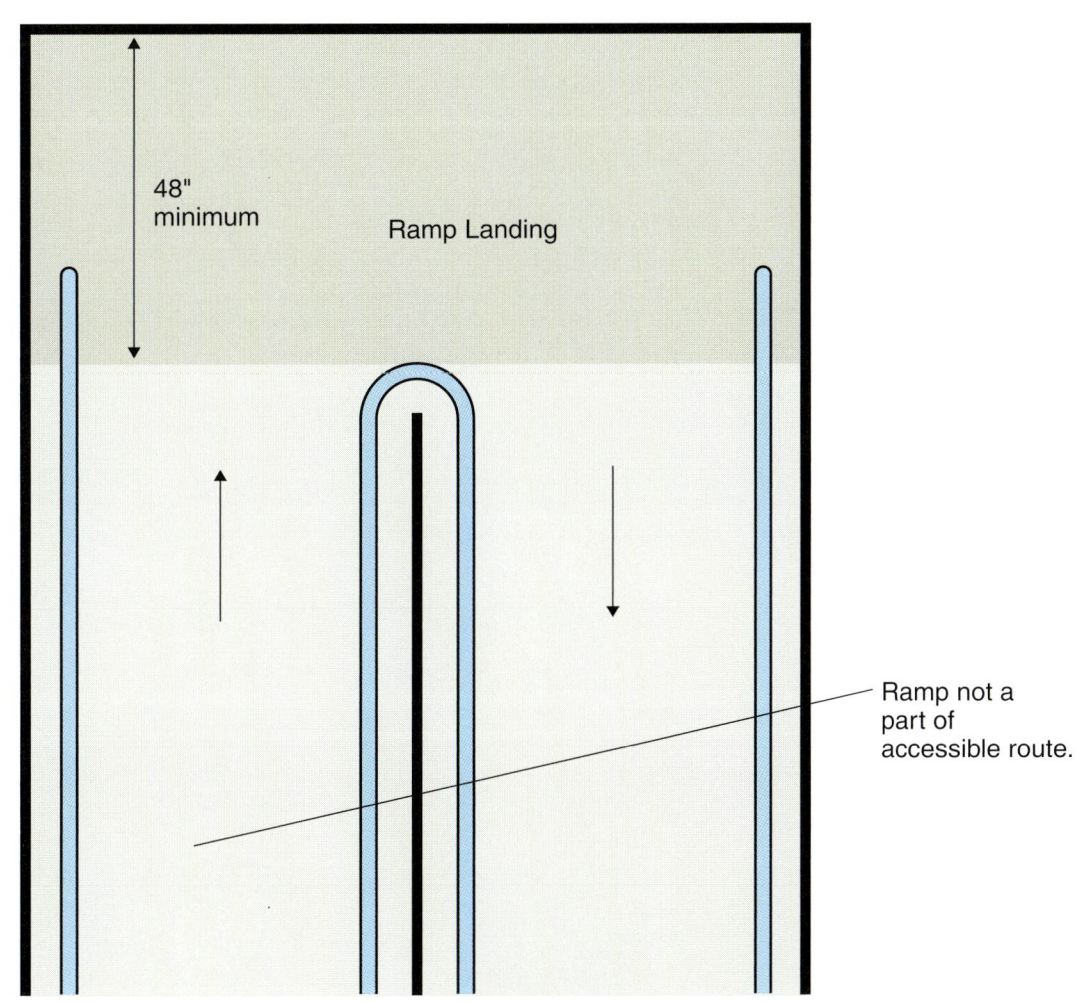

Ramp Landings

1010.6.3 continued

CHANGE SIGNIFICANCE. The minimum 60-inch dimension for ramp landings in the general provisions of the code is based primarily on wheelchair use. This extended length provides sufficient distance to stop and rest if necessary, and it allows enough space to negotiate a turn with minimal effort. Where the ramp is not a portion of an accessible route, a 60-inch length is not considered necessary. The revised minimum landing length of 48 inches has been deemed adequate for egress purposes.

1010.9, 1010.9.1, 1010.9.2

Edge Protection at Ramps

CHANGE TYPE. Addition

CHANGE SUMMARY. An extended floor surface is now required where handrails are used as the method of edge protection for the sides of ramp runs and landings. **(E54-03/04; E54-04/05)**

2006 CODE: 1010.9 Edge Protection. Edge protection complying with Section 1010.9.1 or 1010.9.2 shall be provided on each side of ramp runs and at each side of ramp landings.

Exceptions:
1. through 3. (no change to text)

~~**1010.9.1 Railings.** A rail shall be mounted below the handrail 17 inches to 19 inches (432 mm to 483 mm) above the ramp or landing surface.~~

~~**1010.9.2**~~ **1010.9.1 Curb, Rail, Wall, or Barrier.** A curb, rail, wall, or barrier shall be provided that prevents the passage of a 4-inch-diameter (102 mm) sphere, where any portion of the sphere is within 4 inches (102 mm) of the floor or ground surface.

1010.9.2 Extended Floor or Ground Surface. The floor or ground surface of the ramp run or landing shall extend 12 inches (305 mm) minimum beyond the inside face of a handrail complying with Section 1012.

CHANGE SIGNIFICANCE. The code has previously permitted two methods of edge protection at ramp runs and ramp landings where the ramp is used as a component of a means of egress. These methods included the installation of railings or the use of curbs or barriers.

The accessibility provisions of ICC A117.1 allow the use of curbs and barriers as well, but where handrails are used, an extended floor surface is also required. ICC A117.1 requirements for extended floor

1010.9, 1010.9.1, 1010.9.2 continues

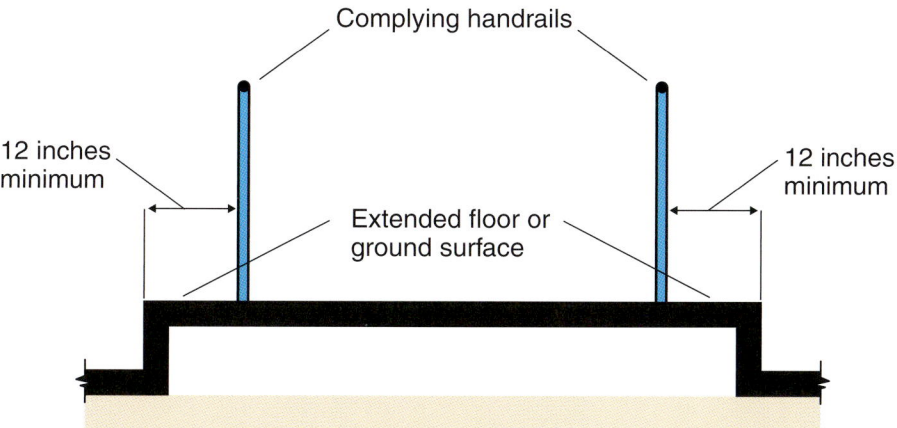

Ramp Edge Protection

Complying handrails

12 inches minimum

12 inches minimum

Extended floor or ground surface

1010.9, 1010.9.1, 1010.9.2
continued

surfaces have now been added to Section 1010.9.2 so that the provisions in this section correlate with the accessible ramp provisions of ICC A117.1.

Previously, a handrail provided with a mid-height rail was permitted by the International Codes for edge protection along the sides of ramps and their associated landings. However, with no extension of the floor surface beyond the rail, the intermediate rail would not always stop a crutch tip or the front wheels of a wheelchair from leaving the ramp surface. The new requirements for a minimum extension of 12 inches beyond the inside face of the handrail addresses these concerns.

1013.3
Guard Opening Limitations for Group R-2 Occupancies

CHANGE TYPE.   Addition

CHANGE SUMMARY.   In individual dwelling and sleeping units of Groups R-2 and R-3 occupancies, the maximum permitted opening between intermediate rails or ornamental enclosures in required guards at open sides of stairs has been increased to permit a sphere 4⅜ inches in diameter versus the prior limitation of a 4-inch sphere. **(E62-04/05)**

2006 CODE: **1012̲3.3 Opening Limitations.**   Open guards shall have balusters or ornamental patterns such that a 4-inch-diameter (102 mm) sphere cannot pass through any opening up to a height of 34 inches (864 mm). From a height of 34 inches (864 mm) to 42 inches (1067 mm) above the adjacent walking surfaces, a sphere 8 inches (203 mm) in diameter shall not pass.

Exceptions:
1. **through 4.** (no change to text)
5. Within individual dwelling and sleeping units in Group R-2 and R-3 occupancies, openings for required guards on the sides of stair treads shall not allow a sphere of 4.375 inches (111 mm) to pass through.

1013.3 continues

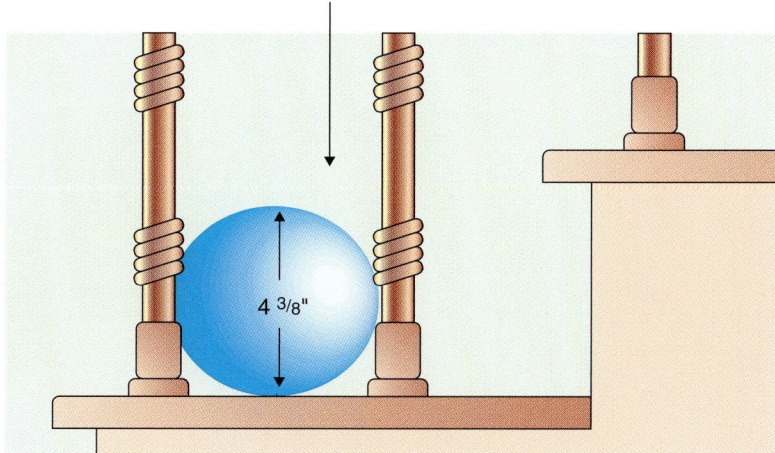

Applicable to stairs within sleeping units and dwelling units of Group R-2 and Group R-3 occupancies.

4 3/8"

Guard Openings

1013.3 continued **CHANGE SIGNIFICANCE.** Previously, the IBC limited the opening size in required guards on the sides of stairs to only those configurations where a 4-inch sphere could not pass through. The ⅜-inch increase in opening size now allows for the placement of only two balusters at each tread, based on a 10-inch tread depth, rather than three balusters. Because a tread depth of 10 inches is permitted in only limited residential occupancies, the increased opening size is appropriate only within dwelling and sleeping units of Group R-2 and R-3 occupancies.

The general provisions governing maximum opening size for other guards remain unchanged, limiting the openings created by intermediate rails or ornamental closures to those where a 4-inch sphere is unable to pass through. The limited increase to 4⅜ inches along the open sides of stairs was deemed acceptable because the primary concern associated with the opening size is preventing infants from crawling through guards. On stairs, this is a minor concern versus that of infants falling down the stairs.

CHANGE TYPE.   Modification

CHANGE SUMMARY.   The means of egress is now permitted to pass through a stockroom serving a Group M occupancy, provided four specified conditions are met, addressing hazard level, alternative egress routes, door hardware, and aisle configuration. **(E63-03/04; E67-04/05; E68-04/05)**

2006 CODE: **101~~3~~4.2 Egress through Intervening Spaces.**   <u>Egress through intervening spaces shall comply with this section.</u>

1. Egress from a room or space shall not pass through adjoining or intervening rooms or areas, except where such adjoining rooms or areas are accessory to the area served; are not a high-hazard occupancy; and provide a discernible path of egress travel to an exit *(balance of this paragraph and prior Exception 1 relocated to Exceptions 2, 3, and 4 to Item 2).*

   Exceptions: ~~2.~~ Means of egress are not prohibited through adjoining or intervening rooms or spaces in a Group H<u>, S, or</u>

1014.2 continues

1014.2

Egress through Intervening Spaces

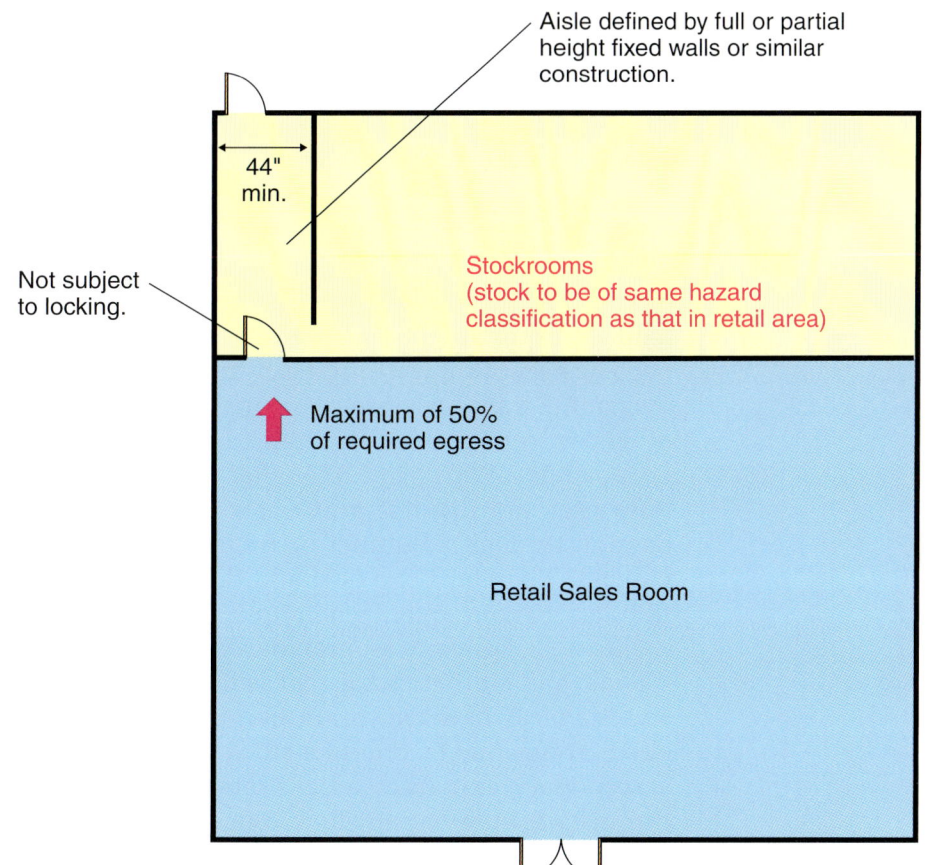

Egress Through Stockrooms Serving Group M Occupancies

1014.2 continued

F occupancy when the adjoining or intervening rooms or spaces are the same or a lesser hazard occupancy group.

2. Egress shall not pass through kitchens, storage rooms, closets, or spaces used for similar purposes.

Exceptions:
1. Means of egress are not prohibited through a kitchen area serving adjoining rooms constituting part of the same dwelling unit or sleeping unit.

2. Means of egress are not prohibited through stockrooms in Group M occupancies when all of the following are met:
   2.1 The stock is of the same hazard classification as that found in the main retail area;
   2.2 Not more than 50% of the exit access is through the stockroom;
   2.3 The stockroom is not subject to locking from the egress side; and
   2.4 There is a demarcated, minimum 44-inch-wide (1118 mm) aisle defined by full or partial height fixed walls or similar construction that will maintain the required width and lead directly from the retail area to the exit without obstructions.

3. An exit access shall not pass through a room that can be locked to prevent egress.

4. Means of egress from dwelling units or sleeping areas shall not lead through other sleeping areas, toilet rooms, or bathrooms.

CHANGE SIGNIFICANCE. Item 1: Provisions previously in Exception 2, which now appear in the exception to Item 1, allowed egress thorough Group H occupancies, where such occupancies were the same or a lesser hazard use as compared to the space where egress originated. The addition of Groups S and F occupancies to this exception has no consequence and apparently resulted from a misunderstanding of the intended purpose of the exception.

The exception was originally added as a permissible allowance for overriding the restriction in the main paragraph that prohibited means of egress from passing through adjoining or intervening high-hazard occupancies. With respect to other occupancy classifications, the code does not, and did not previously, restrict the means of egress from passing through adjoining or intervening rooms of such other occupancies. Therefore, the new addition of Groups S and F in the exception has no consequence on application of the code.

Item 2: Egress was previously prohibited through stock rooms, closets, and similar storage areas because of the high potential for obstructions. A new exception permits one of the most commonly desired arrangements, that of a stock room serving as part of the path to a secondary exit from a sales area. Egress is now allowed to pass through an intervening stock room, provided that several conditions are satisfied.

First, the allowance is limited to stock rooms serving Group M occupancies, and the stock must be of the same hazard classification as the merchandise in the sales area. This restriction ensures that the combustibility of the building's contents will be consistent between the two areas.

Second, the egress capacity permitted through the stock room is limited to one-half of that required for the sales area. Thereby, at least 50% of the required means of egress must be available from the sales area without passing through the stock room. In addition, locking devices capable of obstructing egress are prohibited on any door connecting the sales area to the stockroom providing the secondary means of egress.

In an effort to minimize the potential for obstructions in the egress path through the stock room, a substantial and permanent method of defining the egress aisle is required. The aisle must be a minimum of 44 inches in width, and in no case less than that required on the basis of the occupant load served. Full- or partial-height walls or similar construction that permanently establishes the egress path must be provided in a manner to maintain the minimum required unobstructed width. It is not adequate or permissible to simply designate the required egress path by marking the floor surface.

1014.2.1

Egress through Adjoining Tenant Spaces

CHANGE TYPE.   Addition

CHANGE SUMMARY.   Egress from a small tenant space through a larger adjoining tenant space is now permitted where the tenants have comparable uses, the egress path is discernable, and locking devices do not restrict egress travel. **(E70-04/05)**

2006 CODE: 1013̲4.2.1 Multiple Tenants.   Where more than one tenant occupies any one floor of a building or structure, each tenant space, dwelling unit, and sleeping unit shall be provided with access to the required exits without passing through adjacent tenant spaces, dwelling units, and sleeping units.

> **Exception:** Means of egress shall not be prohibited through adjoining tenant space where such rooms or spaces occupy less than 10% of the area of the tenant space through which they pass; are the same or similar occupancy group; a discernable path of egress travel to an exit is provided; and the means of egress into the adjoining space is not subject to locking from the egress side. A required means of egress serving the larger tenant space shall not pass through the smaller tenant space or spaces.

CHANGE SIGNIFICANCE.   It is not uncommon for a large tenant space, such as a retail department/grocery store, to also house one or more smaller tenants, such as a bank branch or a fast-food restaurant. Previous code text required such small tenants to have a means of egress that is independent of the larger tenant, typically resulting in

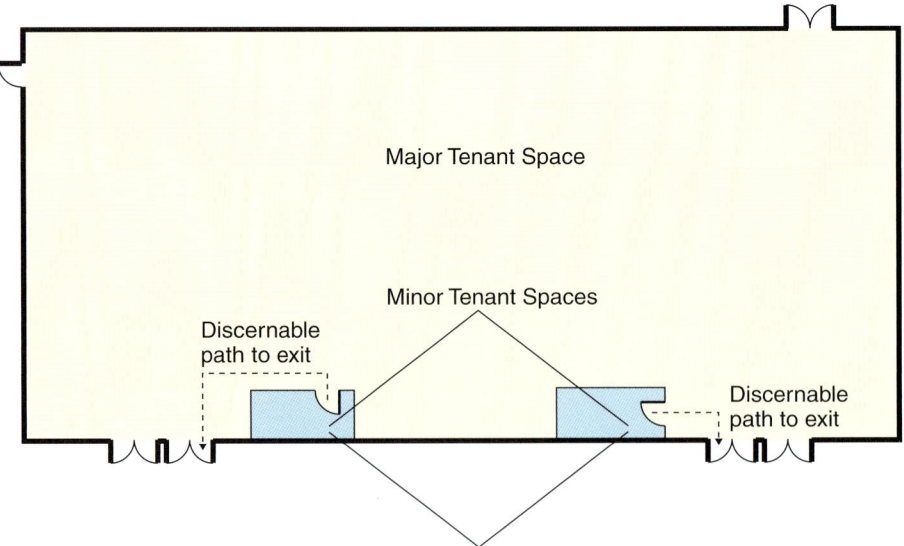

• Floor area limited to 10% of major tenant space

• Means of egress not subject to locking from major tenant side

Egress Through Adjoining Tenant Spaces

additional exit doors directly to the exterior of the building. A new exception allows for egress from such small tenants through a larger tenant space, provided several conditions are met.

The small tenant spaces are limited in floor area to 10% of the area of the major tenant space, restricting use of the exception to relatively small spaces. The hazard level of the tenants must all be relatively consistent, as in the case of a Group M occupancy and a Group B occupancy. In addition, the path of travel through the adjoining tenant space must be discernable, unobstructed, and not subject to locking from the egress side.

1014.3

Common Path of Egress Travel in Group R-2 Occupancies

CHANGE TYPE.   Modification

CHANGE SUMMARY.   The maximum permitted length of the common path of egress travel in Group R-2 occupancies has been increased from 75 feet to 125 feet. **(E72-04/05; E74-04/05)**

2006 CODE: 1013̶4.3 Common Path of Egress Travel.   In occupancies other than Groups H-1, H-2, and H-3, the common path of egress travel shall not exceed 75 feet (22 860 mm). In occupancies in Groups H-1, H-2, and H-3, the common path of egress travel shall not exceed 25 feet (7620 mm). For common path of egress travel in Group A occupancies having fixed seating, see Section 1025.8.

Exceptions:
1. through 3. (no change to text)
4. The length of a common path of egress travel in a Group R-2 occupancy shall not be more than 125 feet (38 100 mm), provided that the building is protected throughout with an approved automatic sprinkler system in accordance with Section 903.3.1.1.

CHANGE SIGNIFICANCE.   In today's residential housing market, larger dwelling units are becoming much more prevalent. The increase in the permitted length of a common path of egress travel in a Group R-2 occupancy recognizes the extended travel distances that

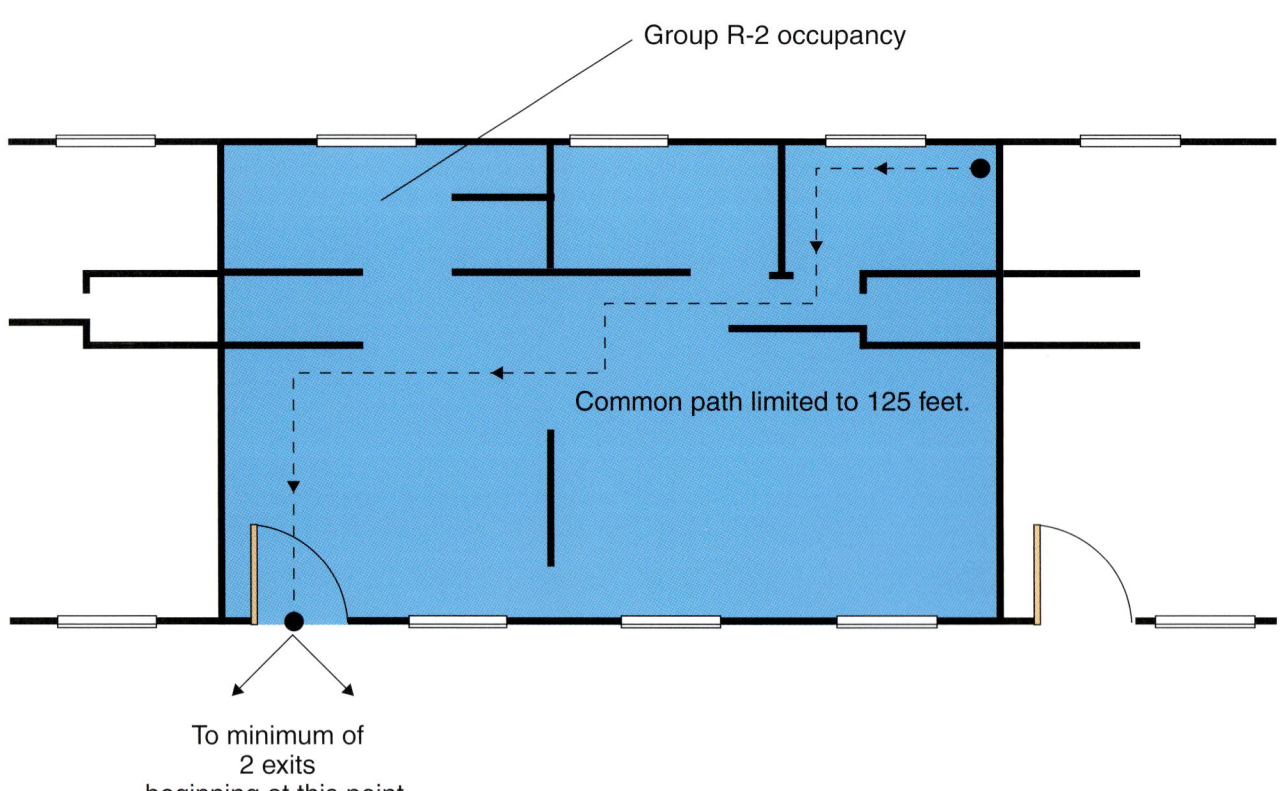

Group R-2 occupancy

Common path limited to 125 feet.

To minimum of
2 exits
beginning at this point

Common Path of Egress Travel in Group R-2

are created, with typically no additional occupant load expected. A maximum distance of 125 feet is now permitted for travel in the exit access before separate and distinct paths of egress travel to two exits are available.

The effect of this code change is bounded by Section 1015.1, which regulates the minimum required number of exits or exit access doorways. In addition to complying with the common path provisions, single-exit Group R-2 occupancies are limited to a maximum occupant load of 10. On the basis of a density factor of 200 square feet per occupant, a single-exit dwelling unit or sleeping unit would be limited to 2000 square feet regardless of the common path limit. The increased allowance for common path of travel will permit a previously unavailable option to have a single means of egress in some larger dwelling and sleeping units that do not exceed 2000 square feet in area.

The use of the exception is limited to only those Group R-2 occupancies that are sprinklered in accordance with Section 903.3.1.1, which references NFPA 13. The proponent of this code change only requested consideration of the allowance for buildings protected with NFPA 13 compliant sprinkler systems, so the increased common path does not apply to buildings protected with NFPA 13R sprinkler systems.

A cross-reference to Section 1025.8 was also added to address common path of egress travel in Group A occupancies. Where fixed assembly seating is provided, the provisions of Section 1025.8 apply, rather than those of Section 1014.3.

1014.4.2

Aisle Accessways in Group M Occupancies

CHANGE TYPE.   Addition

CHANGE SUMMARY.   Provisions have been added to address aisle accessways in Group M occupancies, mandating a minimum clear width of 30 inches on at least one side of each element within a merchandise pad. **(E75-04/05)**

2006 CODE: **1014.4.2 Aisle Accessways in Group M.**   An aisle accessway shall be provided on at least one side of each element within the merchandise pad. The minimum clear width for an aisle accessway not required to be accessible shall be 30 inches (762 mm). The required clear width of the aisle accessway shall be measured perpendicular to the elements and merchandise within the merchandise pad. The 30-inch (762 mm) minimum clear width shall be maintained to provide a path to an adjacent aisle or aisle accessway. The common path of travel shall not exceed 30 feet (9144 mm) from any point in the merchandise pad.

> **Exception:** For areas serving not more than 50 occupants, the common path of travel shall not exceed 75 feet (22 880 mm).

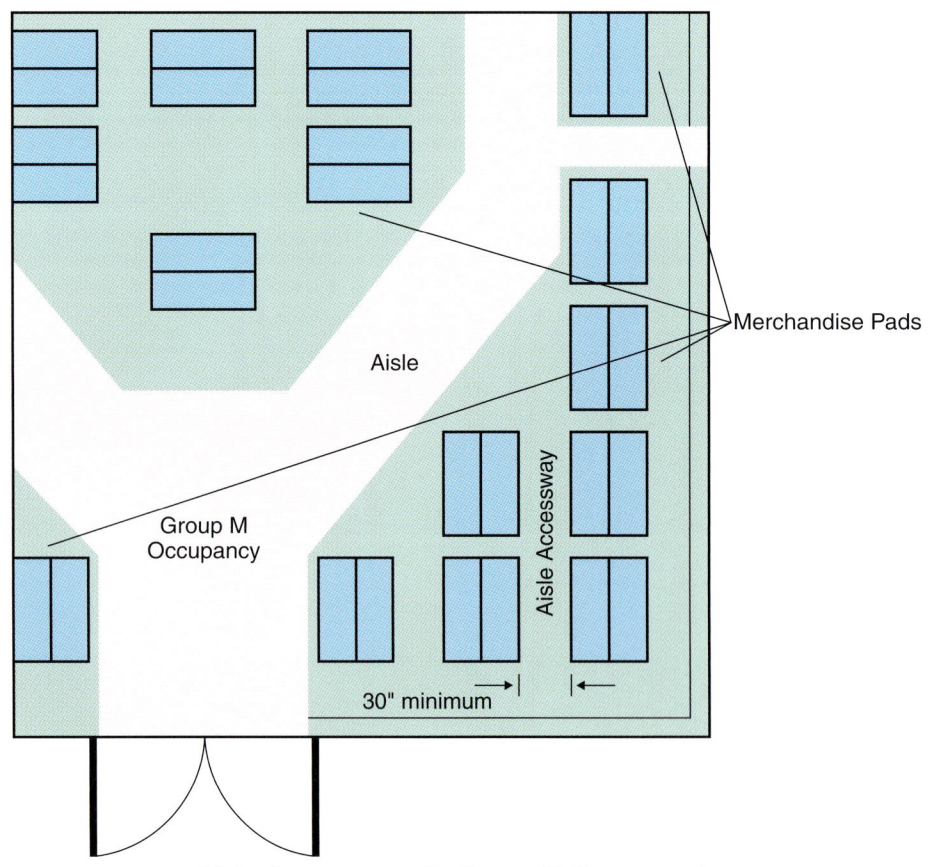

Aisle Accessways in Group M Occupancies

SECTION 1002 DEFINITIONS

AISLE. An exit access component that defines and provides a path of egress travel.

MERCHANDISE PAD. A merchandise pad is an area for display of merchandise surrounded by aisles, permanent fixtures, or walls. Merchandise pads contain elements such as nonfixed and moveable fixtures, cases, racks, counters, and partitions as indicated in Section 105.2 of the *International Building Code* from which customers browse or shop.

CHANGE SIGNIFICANCE. Although aisles in Group M occupancies have long been regulated for width purposes, provisions did not specifically address the unique characteristics of merchandise display areas. This code change adds new provisions addressing aisle accessways, clarifying that the requirements differ somewhat from those regulating aisles. In addition, a definition of merchandise pad has been provided to identify the limits of the aisle accessway requirements.

A merchandise pad is defined as the merchandise display area that contains multiple counters, shelves, racks, and other movable fixtures. Bounded by aisles, permanent fixtures, and walls, the merchandise pad also includes aisle accessways utilized to provide both access to an aisle and circulation throughout the pad area. Every element within a merchandise pad must adjoin a minimum 30-inch-wide aisle accessway on at least one side. Travel within a merchandise pad is limited, with a maximum common path of travel of 30 feet. The common path of travel limitation is extended to 75 feet in those areas serving a maximum occupant load of 50.

A merchandise pad in a retail store.

Table 1015.1

Single Means of Egress from Day Care Uses

CHANGE TYPE. Modification

CHANGE SUMMARY. The maximum occupant load permitted for a Group E day care occupancy with a single means of egress has been reduced from 50 to 10 occupants. Additionally, thresholds limiting buildings and spaces with a single means of egress or exit have been reduced by one occupant for correlation with other code sections with similar thresholds. **(G55-04/05; E1-04/05; E10-04/05; E108-04/05)**

2006 CODE:

TABLE 1015~~4~~.1 Spaces with One Means of Egress

Occupancy	Maximum Occupant Load
A, B, E[a], F, M, U	~~50~~ 49
H-1, H-2, H-3	3
H-4, H-5, I-1, I-3, I-4, R	10
S	~~30~~ 29

a. Day care maximum occupant load is 10.

TABLE 101~~9~~8.2 Buildings with One Exit

Occupancy	Maximum Height of Building Above Grade Plane	Maximum Occupants (or Dwelling Units) per Floor and Travel Distance
A, B[d], E[e], F, M, U	1 Story	~~50~~ 49 occupants and 75 feet travel distance
H-2, H-3	1 Story	3 occupants and 25 feet travel distance
H-4, H-5, I, R	1 Story	10 occupants and 75 feet travel distance
S[a]	1 Story	~~30~~ 29 occupants and 100 feet travel distance
B[b], F, M, S[a]	2 Stories	30 occupants and 75 feet travel distance
R-2	2 Stories[c]	4 dwelling units and 50 feet travel distance

For SI: 1 foot = 304.8 mm.

a. For the required number of exits for open parking structures, see Section 1019.1.1.

b. For the required number of exits for air traffic control towers, see Section 412.1 of the *International Building Code.*

c. Buildings classified as Group R-2 equipped throughout with an automatic sprinkler system in accordance with Section 903.3.1.1 or 903.3.1.2 and provided with emergency escape and rescue openings in accordance with Section 1026 shall have a maximum height of three stories above grade plane.

d. Buildings equipped throughout with an automatic sprinkler system in accordance with Section 903.3.1.1 with an occupancy in Group B shall have a maximum travel distance of 100 feet.

e. Day care maximum occupant load is 10.

CHANGE SIGNIFICANCE. Classification of a day care facility varies on the basis of several established factors, resulting in either a Group E or Group I-4 designation. Whereas day care uses considered Group I-4 occupancies previously required a second means of egress where the occupant load exceeded 10, those uses classified as Group E were previously permitted to have a single means of egress up to an occupant load of 50 persons. This code change corrects the inconsistency in favor of setting the threshold for two means of egress for any day care facility at 10 occupants. The change, which is accomplished by adding a footnote to Tables 1015.1 and 1019.2, recognizes that providing only one means of egress for more than 10 individuals in a day care facility, many of whom may be infants or small children, may not be adequate to allow for timely evacuation.

Slight changes also occurred in the maximum occupant-load limits for Group A, B, E, F, M, S, and U occupancies with a single means of egress. The maximum number of occupants permitted in single-exit spaces was decreased by one occupant for each of these occupancy classifications to provide a greater degree of consistency with threshold values used elsewhere in the code. Thresholds of 50 and 30 occupants are commonly utilized in the code as points where code requirements are increased, but the code has been inconsistent in the past with respect to "more than 50" versus "50 or more" and "more than 30" versus "30 or more." Changes to these tables in 2006 correlate the thresholds with the "50 or more" and "30 or more" approaches.

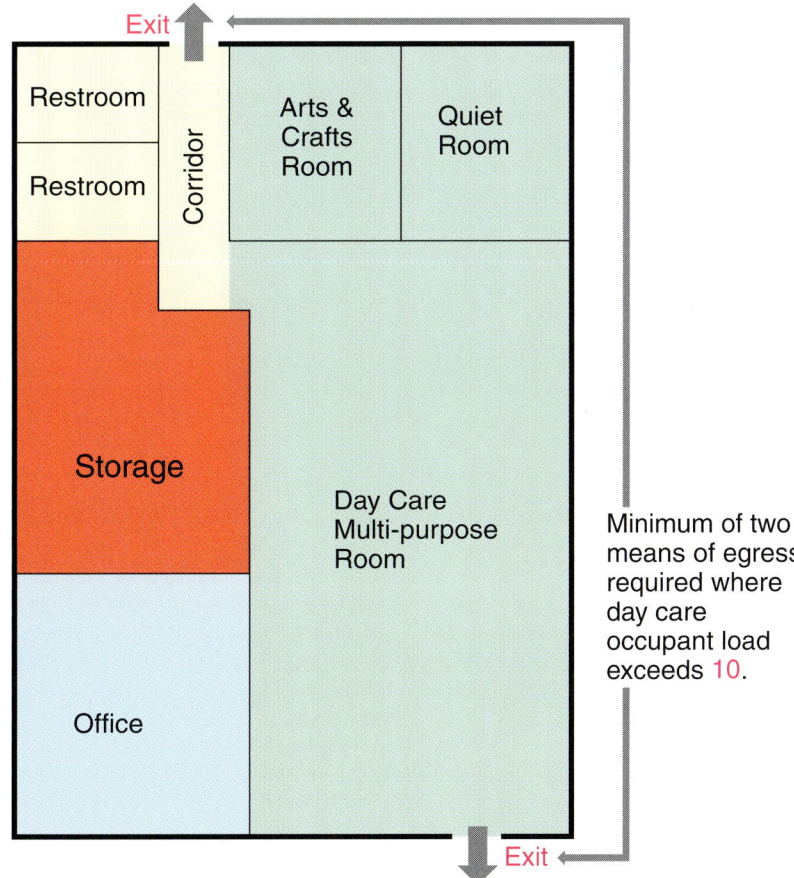

Egress from Day Care Facilities

1015.2.2

Egress Separation of Three or More Exits

CHANGE TYPE. Deletion

CHANGE SUMMARY. In those rooms or buildings where access to three or more exits is required, only two of the doors or doorways are now specifically regulated for exit separation. The performance criteria addressing the location of the additional required exit doors or exit access doorways have been deleted. **(E82-04/05; E84-04/05)**

2006 CODE: 1015.2.2 Three or More Exits or Exit Access Doorways. Where access to three or more exits is required, at least two exit doors or exit access doorways shall be <u>arranged in accordance with the provisions of Section 1015.2.1.</u> ~~placed a distance apart equal to not less than one-half of the length of the maximum overall diagonal dimension of the area served measured in a straight line between such exit doors or exit access doorways. Additional exits or exit ac~~

Example:

Given: A fully sprinklered retail sales building with three required exits.

Determine: The required dispersion of the exits.

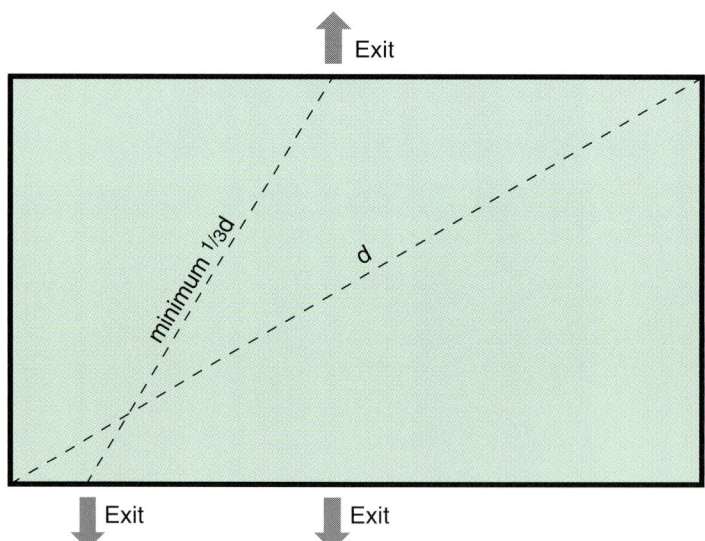

Solution: A minimum of two exit doors must be separated by at least 1/3 the maximum overall diagonal dimension in a sprinklered building. While there is no code requirement regulating placement of the third exit, it should be located so as to provide for a third distinct and independent point of egress.

Separation of Three or More Exits

~~cess doorways shall be arranged a reasonable distance apart so that if one becomes blocked, the others will be available.~~

> ~~**Exception:** Where a building is equipped throughout with an automatic sprinkler system in accordance with Section 903.3.1.1 or 903.3.1.2, the separation distance of at least two of the exit doors or exit access doorways shall not be less than one third of the length of the maximum overall diagonal dimension of the area served.~~

CHANGE SIGNIFICANCE. Where two means of egress are required, they must be adequately separated from each other to reduce the chance that both will be blocked by the same incident. The same criteria apply where three or more means of egress are mandated: at least two exit doors or exit access doorways must be remotely located.

Previously, the code attempted to regulate the placement of the third required exit door or access doorway, where applicable, as well as any additional required doors or doorways. The former requirement for separation of a third exit, fourth exit, etc. was deemed to be vague and unenforceable because it referred to a "reasonable" separation distance. It was also reasoned that spaces requiring 3 or more exits or exit accesses would be relatively large, with separations dictated in part by limits on exit travel distance and common path of travel. In such spaces, the probability of an event that would block more than one exit or exit access is reduced.

With the deletion of text addressing additional exit doors and exit access doorways, the code is now silent as to their required location; consequently, it is now up to the design professional to select suitable locations based on satisfying exit travel distance and common path of travel limitations.

Additional changes to the text, including deletion of the exception, eliminated redundancy by simply referring to the provisions of Section 1015.2.1 addressing the separation of two required means of egress.

1020.1

Unenclosed Interior Exit Stairways

CHANGE TYPE. Modification

CHANGE SUMMARY. A single unenclosed exit stairway is now permitted from a floor level below the exit discharge level, provided the stairway serves a maximum occupant load of nine and interconnects no other floors. Additional changes to this section are intended to clarify previous provisions. **(E90-03/04; E91-03/04; FS2-04/05; FS124-04/05; E106-04/05; E107-04/05)**

2006 CODE: 102019.1 Enclosures Required. Interior exit stairways and interior exit ramps shall be enclosed with fire barriers constructed in accordance with Section 706 of the *International Building Code* or horizontal assemblies constructed in accordance with Section 711 of the *International Building Code, or both.* Exit enclosures shall have a fire-resistance rating of not less than 2 hours where connecting four stories or more and not less than 1 hour where connecting less than four stories. The number of stories connected by the shaft exit enclosure shall include any basements but not any mezzanines. An exit enclosure shall not be used for any purpose other than means of egress.

Exceptions:
1. In all occupancies, other than Group H and I occupancies, a stairway is not required to be enclosed when the stairway serves serving an occupant load of less than 10 and the stairway complies with either item 1.1 or 1.2. In all cases, the maximum number of connecting open stories shall not exceed two.
   1.1. The stairway is open to not more than one story above the story at the level of exit discharge, or
   1.2. The stairway is open to not more than one story below; the story at the level of exit discharge, but not both.
2. Exits in buildings of Group A-5 where all portions of the means of egress are essentially open to the outside need not be enclosed.
3. Stairways serving and contained within a single residential dwelling unit or sleeping unit in Group R-1, R-2, or R-3 occupancies are not required to be enclosed.

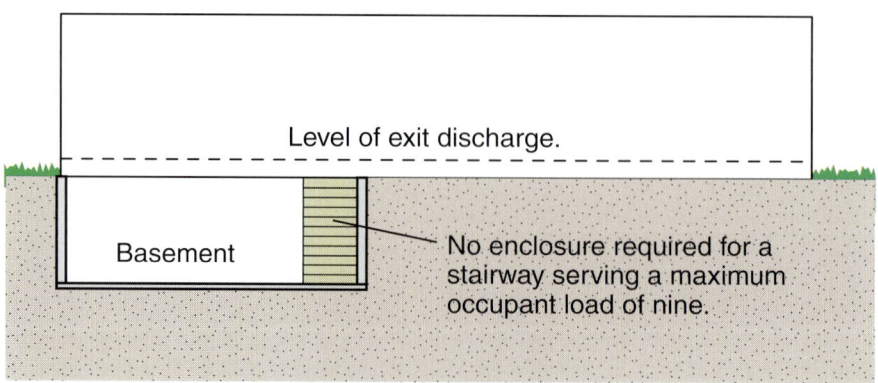

Level of exit discharge.

Basement

No enclosure required for a stairway serving a maximum occupant load of nine.

Unenclosed Stairway

4. Stairways that are not a required means of egress element are not required to be enclosed where such stairways comply with Section 707.2 of the *International Building Code.*

5. Stairways in open parking structures that serve only the parking structure are not required to be enclosed.

6. Stairways in Group I-3 occupancies, as provided for in Section 408.3.6 of the *International Building Code,* are not required to be enclosed.

7. Means of egress stairways as required by Section 410.5.3 of the *International Building Code* are not required to be enclosed.

8. In other than Group H and I occupancies, a maximum of 50 percent of egress stairways serving one adjacent floor are not required to be enclosed, provided at least two means of egress are provided from both floors served by the unenclosed stairways. <u>Any two such interconnected floors shall not be open to other floors. Unenclosed exit stairways shall be remotely located as required in Section 1015.2.</u>

9. In other than Group H and I occupancies, interior egress stairways serving only the first and second stories of a building equipped throughout with an automatic sprinkler system in accordance with Section 903.3.1.1 are not required to be enclosed, provided at least two means of egress are provided from both floors served by the unenclosed stairways. <u>Such interconnected stories shall not be open to other stories. Unenclosed exit stairways shall be remotely located as required in Section 1015.2.</u>

CHANGE SIGNIFICANCE. As a general rule, all interior exit stairways must be enclosed. Exception 1 permits a single stairway to be unenclosed, provided the occupant load served by the stair does not exceed nine. The use of the exception was previously limited to a stairway between the level of exit discharge and the level directly above. The code change permits a similar condition where the stairway connects the discharge level with the story directly below, providing for consistency in application. It is also clearly stated that only two stories are permitted to be interconnected by an unenclosed stairway.

A modification to Exception 3 now makes it applicable to dwelling units in Group R-1 occupancies. Previously, the ability to create unenclosed stairways within units classified as Group R-1 occupancies was limited under this exception to only sleeping units. The code change provides for consistency with the allowances for Groups R-2 and R-3.

Exceptions 8 and 9 have both been modified in order to clarify their application. Where such unenclosed stairways are utilized as portions of the means-of-egress system, they must be remotely located to lessen the possibility that both will be blocked by the same fire incident. Because the egress-remoteness provisions of Section 1015.2 specifically address only the required separation between exit doors and exit-access doorways, the new text clarifies that the concept also applies to unenclosed stairways.

1020.1.7.1

Egress from Smokeproof Enclosures

CHANGE TYPE. Addition

CHANGE SUMMARY. Smokeproof enclosures and pressurized stairways are now permitted to exit through interior areas or vestibules located on the level of exit discharge, under the limitations imposed by Section 1024. **(E114-04/05)**

2006 CODE: **1020.1.7.1** ~~1019.1.8.1~~ **Enclosure Exit.** A smokeproof enclosure or pressurized stairway shall exit into a public way or into an exit passageway, yard, or open space having direct access to a public way. The exit passageway shall be without other openings and shall be separated from the remainder of the building by 2-hour fire-resistance-rated construction.

Exceptions:

1. and 2. (no change to text)

3. <u>A smokeproof enclosure or pressurized stairway shall be permitted to egress through areas on the level of discharge or vestibules as permitted by Section 1024.</u>

CHANGE SIGNIFICANCE. As a general requirement, all smokeproof enclosures and pressurized stairways required in high-rise or underground buildings must exit directly into a public way or into an exit passageway, yard, or open space having direct access to a public way. The continuity and protection provided by the smokeproof enclosure or pressurized stairway in such cases extends to the exterior at grade level. This new exception recognizes two allowances for discontinu-

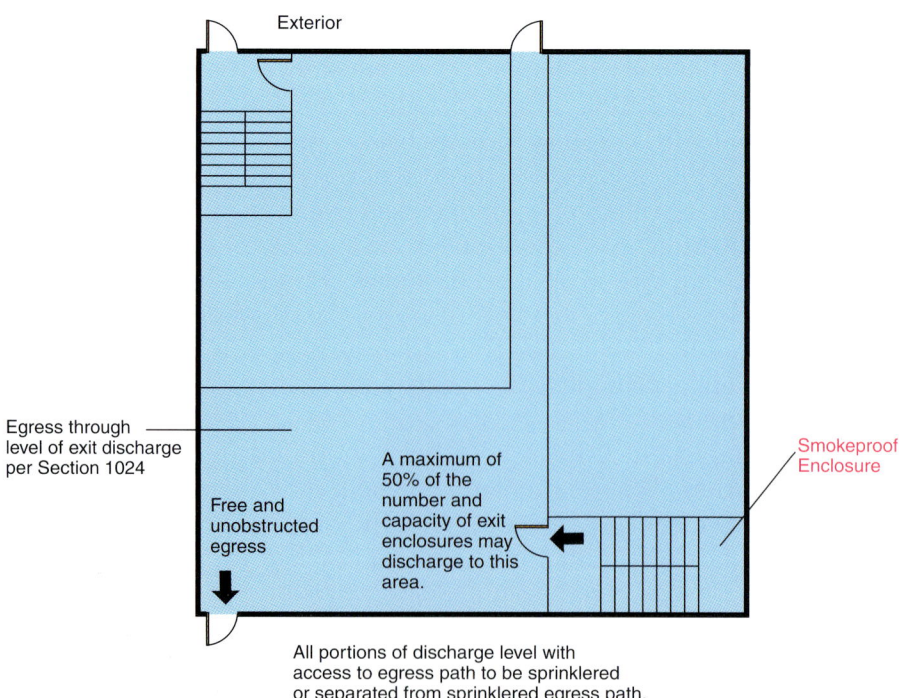

Exterior

Egress through level of exit discharge per Section 1024

Free and unobstructed egress

A maximum of 50% of the number and capacity of exit enclosures may discharge to this area.

Smokeproof Enclosure

All portions of discharge level with access to egress path to be sprinklered or separated from sprinklered egress path.

Egress from Smokeproof Enclosures

ing the protection at a point inside the building, similar to those permitted for other vertical exit enclosures.

The provisions of Section 1024 permit a maximum of 50% of the number and capacity of exit enclosures to egress through areas on the level of exit discharge under specific conditions. Similar allowances are also provided for egress through complying vestibules. The new exception to Section 1020.1.7.1 allows up to 50% of the smokeproof enclosures or pressurized stairways to egress in the same fashion. The extent of the pressurization or smokeproof measures would be limited to within the enclosure and need not continue into the discharge level or vestibule.

1025.3

Egress from Group A Occupancies

CHANGE TYPE.   Clarification

CHANGE SUMMARY.   The method for providing egress from larger assembly spaces has been clarified to indicate that the required means of egress, other than the main exit, can include egress components other than "exits." **(E105-03/04)**

2006 CODE: 1025.3 Assembly Other Exits.   In addition to having access to a main exit, each level of an occupancy in Group A having an occupant load of greater than 300 shall be provided with additional ~~exits~~ means of egress that shall provide an egress capacity for at least one-half of the total occupant load served by that level and comply with Section 1015.2.

Exception: (no change to text)

CHANGE SIGNIFICANCE.   Where a Group A occupancy has an occupant load exceeding 300, access to a main exit must be provided. Additional exits are also mandated; however, it is intended that exit-access components leading to exits also be acceptable. The change makes this clear.

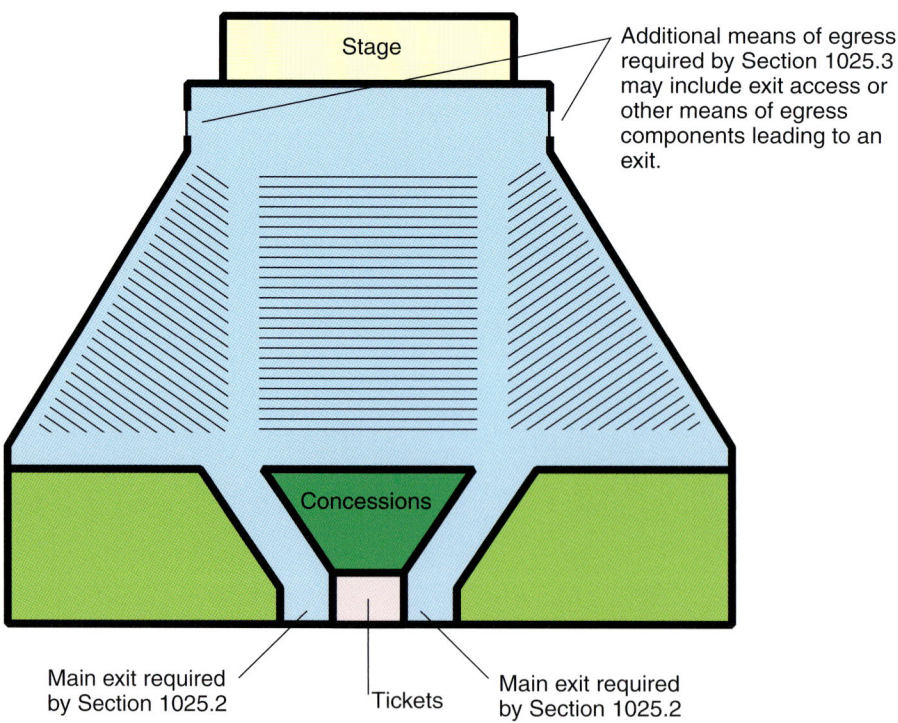

Egress from Group A Occupancies

1028.2

Reliability—Maintenance of the Means of Egress

CHANGE TYPE. Modification

CHANGE SUMMARY. An allowance has been provided to permit the means of egress in unoccupied areas to be secured. **(F171-04/05)**

2006 CODE: 1028̶7.2 Reliability. Required exit accesses, exits or exit discharges shall be continuously maintained free from obstructions or impediments to full instant use in the case of fire or other emergency <u>when the areas served by such exits are occupied.</u> Security devices affecting means of egress shall be subject to approval of the fire code official.

CHANGE SIGNIFICANCE. In the previous edition, the two sentences in this section appeared to conflict because some types of security devices could render the means of egress unusable. The revised text provides a needed clarification that recognizes the common practice of owners securing their premises when they are unoccupied.

The IFC now recognizes the common practices of securing means of egress in unoccupied areas of buildings. (Photo courtesy of Rich Pearson.)

1028.4

Exit Signs— Maintenance

Objects or signage that confuse or impair visibility or identification of an exit are prohibited.

CHANGE TYPE. Addition

CHANGE SUMMARY. A section has been added to require that exit signs in existing buildings be installed and maintained as required for new construction. **(F173-04/05)**

2006 CODE: **1028.4 Exit Signs.**   Exit signs shall be installed and maintained in accordance with Section 1011. Decorations, furnishings, equipment or adjacent signage that impairs the visibility of exit signs, creates confusion, or prevents identification of the exit shall not be allowed.

CHANGE SIGNIFICANCE. Because exit signs are regarded as important for life-safety, the code has been modified to require that signage in existing buildings must comply with code requirements that were previously applicable only to new construction.

In addition, signs in existing buildings are now specifically required to be isolated from clutter and obstructions in the form of drapes, decorations, partitions, and other signs that may prevent or distract attention. By including a requirement in Section 1028, fire inspectors now have an appropriate requirement for enforcing exit sign maintenance in existing buildings.

1028.7

Testing and Maintenance— Communication Systems for Areas of Refuge

CHANGE TYPE.   Addition

CHANGE SUMMARY.   The code now requires annual inspection and testing of communications systems provided for areas of refuge. **(E30-04/05)**

2006 CODE: **1028.7 Testing and Maintenance.** All two-way communication systems for areas of refuge, shall be inspected and tested on a yearly basis to verify that all components are operational. When required, the tests shall be conducted in the presence of the fire code official.

CHANGE SIGNIFICANCE.   Neither the IBC nor the IFC previously included a requirement that communications systems provided for areas of refuge be tested on a regular basis. Perhaps this would have been accomplished as part of a fire alarm system test if the communication system was routed through a fire alarm control panel, but arranging the communication system in this manner has never been required.

Since area of refuge communications systems are life-safety systems, it is appropriate that the code requires testing to verify operational status.

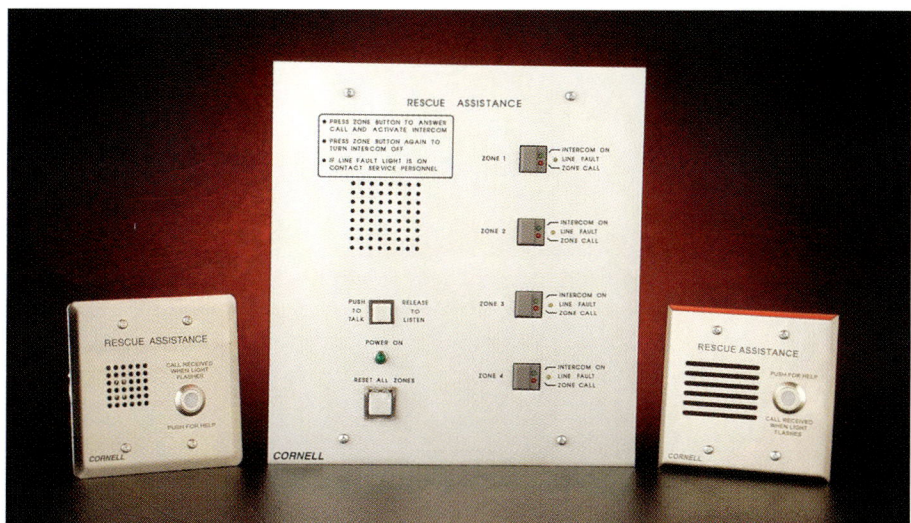

Communication systems in areas of refuge are subject to annual inspection and testing. (Photos courtesy of Cornell Communications Incorporated.)

PART 4

Special Processes and Uses

Chapters 11 Through 26

- **Chapter 11** Aviation Facilities No changes addressed
- **Chapter 12** Dry Cleaning No changes addressed
- **Chapter 13** Combustible Dusty-Producing Operations No changes addressed
- **Chapter 14** Fire Safety During Construction and Demolition No changes addressed
- **Chapter 15** Flammable Finishes
- **Chapter 16** Fruit and Crop Ripening No changes addressed
- **Chapter 17** Fumigation and Thermal Insecticidal Fogging No changes addressed
- **Chapter 18** Semiconductor Fabrication Facilities

- **Chapter 19** Lumber Yards and Woodworking Facilities No changes addressed
- **Chapter 20** Manufacture of Organic Coatings No changes addressed
- **Chapter 21** Industrial Ovens No changes addressed
- **Chapter 22** Motor Fuel-Dispensing Facilities and Repair Garages
- **Chapter 23** High-Piled Combustible Storage
- **Chapter 24** Tents, Canopies and Other Membrane Structures No changes addressed
- **Chapter 25** Tire Rebuilding and Tire Storage No changes addressed
- **Chapter 26** Welding and Other Hot Work No changes addressed

174

Chapters 11 through 26 of the *International Fire Code* establish the minimum requirements for special processes and uses. These processes or uses involve a variety of combustible materials and hazardous materials and are often unique to particular applications, such as semiconductor fabrication facilities or industrial ovens. Certain activities, such as aviation facilities or motor-fuel dispensing facilities, often require the storage and handling of thousands of gallons of flammable or combustible liquids. ■

CHAPTER 15
Flammable Finishes

1504.2
Location of Spray-Finishing Operations

1504.7.6
Ventilation Termination Point

TABLE 1805.2.2 AND SECTION 1805.2.3.5
Maximum Quantities of Hazardous Production Materials (HPMs) at a Workstation

2205.6
Warning Signs

TABLE 2206.2.3
Minimum Separation Requirements for Aboveground Tanks

2209.3.2 AND 2211
Location of Dispensing Operations and Equipment for Hydrogen Motor Fuel

2308.2.1
Plastic Pallets and Shelves in Rack Storage

Chapter 15

Flammable Finishes

CHANGE TYPE.   Modification

CHANGE SUMMARY.   Chapter 15 has been completely reorganized and updated in an effort to clarify the intent of provisions in the prior editions of the code. **(F179-04/05)**

2006 CODE: This chapter was revised in its entirety. For complete code text, see the 2006 *International Fire Code*.

CHANGE SIGNIFICANCE.   Chapter 15 has been completely reorganized to update and editorially clarify previously existing provisions. Some new text was added to improve flow and clarity, but the intent of the proposal that accomplished the rewrite, F179-04/05, was to make no technical changes.

Two technical changes were made to Chapter 15 by other proposals between the 2003 and 2006 editions, modifying Sections 1504.2 and 1504.7.6. These changes are discussed individually after this general discussion of the Chapter 15 rewrite.

Some notable aspects of the Chapter 15 rewrite include:

1. **Location of spray finishing.** In other than Groups A, E, I, and R occupancies, Section 1504.2 (previously 1504.1) requires spray finishing operations to be conducted in spray booths or spray rooms, except as permitted in approved spraying spaces. Previous code editions referenced spray booths, spray rooms, or "limited" spraying spaces, but spraying spaces in the general sense were not recognized in this section.

   This lack of recognition in what is now Section 1504.2 created confusion for some because spraying spaces, as referenced

The requirements for spray finishing, powder coating, manufacturing of reinforced plastics, and other topics regulated by Chapter 15 have been reorganized.

in Chapter 15, otherwise have an appearance of being parallel to spray booths and spraying rooms as permissible places where spray finishing operations may take place (because spraying spaces have their own code section, 1504.3.3 [1504.1.3 in the 2003 edition], which is at a section level parallel to spray booths and spraying rooms).

To provide additional clarity, Section 1504.2 now lists spraying spaces as an alternative location for spray finishing operations, when approved. In addition, the term "spraying space" is now defined in Section 1502.1.

In reading Sections 1504.3.3 and 1504.4, one might conclude that the allowance of spray finishing operations to be conducted in spraying spaces has created a loophole in the code permitting spray finishing operations to be conducted without requiring an approved automatic fire-extinguishing system. Section 1504.4 implies this because spraying spaces are not listed as areas requiring fire protection; however, the IBC fills this void by requiring such protection in Section 416.4.

2. **Limited spraying spaces.** With spraying spaces now being recognized in a general sense as permissible locations for spray finishing operations in Section 1504.2, the allowance for spray finishing in "limited spraying spaces" has been moved from the main paragraph in Section 1504.2 to an exception (Exception 2). Unlike spraying spaces, limited spraying spaces are not required to comply with many of the regulations that are normally associated with spray finishing operations. Section 1504.9 details applicable regulations, which are minimal because of the small size of these spaces, and a restriction to noncontinuous spraying.

The original intent of the code in recognizing limited spraying spaces as unique spraying areas was to allow occasional priming of a single body panel on a vehicle outside of a

Chapter 15 continues

Chapter 15 continued

spray booth, permitting body shops with a single booth to make more efficient use of booth space. However, provisions in this section have become more broadly applied since they were originally incorporated into Article 45 of the Uniform Fire Code for this reason many years ago.

3. **Vapor and spray areas.** The prior terms "vapor area" and "spray area" have been replaced by a single term, "flammable vapor area." Because the hazards to be mitigated in spray and vapor areas were the same, there was no apparent reason for maintaining separate terms.

4. **Resin application areas.** The code has been revised to clarify that spray booths, spray rooms, and spraying spaces are not required for resin application areas, where flammable resins are used. One might expect to find resin application areas in facilities where fiberglass, cultured marble, and similar products are produced. Section 1504.3, Exception 3, now specifically states that compliance with Section 1509 is permitted in lieu of protection features normally associated with spray finishing operations. Section 1503.2.1.2 grants an additional allowance for resin application areas by not requiring hazardous-location electrical equipment.

   Resin application processes, which often involve spray-applied gel-coat and fiberglass "chopper guns," were granted special treatment in the code because the American Composites Manufacturers Association conducted tests demonstrating the reduced ignition hazard of these operations as compared to other types of flammable finishing.

5. **Automobile undercoating.** Automobile undercoating is no longer covered in a separate section (previously 1508) since the purpose of that section was limited to providing an exception to spray finishing regulations. The provisions are now more suitably located in Exception 1 to Section 1504.2.

The following table cross-references 2003 versus 2006 section numbers to assist code users in finding provisions in the new Chapter 15:

2006 Edition	2003 Edition
Section 1501	**Section 1501**
1501.1	1501.1
1501.2	1501.2
Section 1502	**Section 1502**
1502.1	1502.1
Section 1503	**Section 1503**
1503.1	1503.1
1503.2	1503.2
1503.2.1	1503.2.1
1503.2.1.1	1503.2.1.1
1503.2.1.2	1503.2.1.2
1503.2.1.3	1503.2.1.3
1503.2.1.4	1503.2.1.4
1503.2.1.5	1503.2.1.5
1503.2.1.6	1503.2.1.6

2006 Edition	2003 Edition
1503.2.2	1503.2.2
1503.2.3	1503.2.3
1503.2.4	1503.2.4
1503.2.5	1503.2.5
1503.2.6	1503.2.6, 1507.6.3
1503.2.7	1503.2.7
1503.2.8	1503.2.8
1503.3	1503.3
1503.3.1	1503.3.1
1503.3.2	1503.3.2
1503.3.3	1503.3.3
1503.3.4	1503.3.4
1503.3.5	1503.3.5
1503.3.5.1	1503.3.5.1
1503.3.5.2	1503.3.5.2
1503.3.6	1503.3.6
1503.4	1503.4
1503.4.1	1503.4.1
1503.4.2	1503.4.2
1503.4.3	1503.4.3
1503.4.4	1503.4.4
Section 1504	**Section 1504**
1504.1 (New)	None
1504.2	1504.1 & 1508
1504.3 (New)	None
1504.3.1	1504.1.1
1504.3.1.1	1504.1.1.1
1504.3.2	1504.1.2
1504.3.2.1	1504.1.2.1
1504.3.2.2	1504.1.2.2
1504.3.2.3	1504.1.2.3
1504.3.2.4	1504.1.2.4
1504.3.2.5	1504.1.2.5
1504.3.2.6	1504.1.2.6
1504.3.3	1504.1.3
1504.3.3.1	1504.1.3.1
1504.4	1504.6
1504.4.1	1504.6.4
1504.5 (New)	None
1504.5.1	1504.4
1504.5.2	1504.6.1
1504.6 (New)	None
1504.6.1	1504.7 & 1504.7.2.3
1504.6.1.1	1504.7.1
1504.6.1.2	1504.7.2
1504.6.1.2.1	1504.7.2.1
1504.6.1.2.2	1504.7.2.2
1504.6.2	1504.5
1504.6.2.1	1504.5.1
1504.6.2.2	1504.5.2
1504.6.2.3	1504.5.3
1504.6.2.4	1504.5.4
1504.7	1504.2
1504.7.1	1504.2.1
1504.7.2	1504.2.2
1504.7.3	1504.2.3
1504.7.4	1504.2.4

Chapter 15 continues

Chapter 15 continued

2006 Edition	2003 Edition
1504.7.5	1504.2.5
1504.7.6	1504.2.6
1504.7.7	1504.2.7
1504.7.8	1504.3
1504.7.8.1	1504.3.1
1504.7.8.2	1504.3.2
1504.7.8.3	1504.3.3
1504.7.8.4	1504.3.4
1504.7.8.5	1504.3.5
1504.7.8.6	1504.3.6
1504.7.8.7	1504.3.7
1504.8 (New)	None
1504.8.1	1504.6.2
1504.8.1.1	1504.6.2.1
1504.8.2	1504.6.3
1504.9	1504.1.4
1504.9.1	1504.1.4.1
1504.9.2	1504.1.4.2
1504.9.3	1504.1.4.3
1504.9.4	1504.1.4.4
Section 1505	**Section 1505**
1505.1 (New)	None
1505.2	1505.1
1505.3	1505.3
1505.3.1	1505.3.1
1505.3.2	1505.3.2
1505.3.3	1505.3.3
1505.3.4.	1505.7
1505.3.4.1	1505.7.1
1505.3.4.2	1505.7.2
1505.3.4.3	1505.7.3
1505.4 (New)	None
1505.4.1	1505.6
1505.4.1.1	1505.6.1
1505.4.2	1505.5
1505.5 (New)	None
1505.6 (New)	None
1505.7	1505.2
1505.8	1505.4
1505.9	1505.8
1505.9.1	1505.8.1
1505.9.2	1505.8.2
1505.9.3	1505.8.3
1505.9.4	1505.8.4
1505.9.5	1505.8.5
1505.10	1505.9
1505.10.1	1505.9.1
1505.11	1505.10
Section 1506	**Section 1507**
1506.1	1507.1
1506.2	1507.2
1506.3	1507.2
1506.4	1507.8
1506.4.1	1507.9
1506.4.2	1507.1
1506.5	1507.6
1506.5.1	1507.6.1

2006 Edition	2003 Edition
1506.6 (New)	None
1506.6.1	1507.5
1506.6.2	1507.6.2
1506.6.3	1507.6.3
1506.6.4	1507.6.4
1506.7	1507.4
Section 1507	**Section 1506**
1507.1	1506.1
1507.2	1506.4
1507.3	1506.3
1507.3.1	1506.8
1507.4	1506.12
1507.4.1	1506.7
1507.5 (New)	None
1507.5.1	1506.11
1507.5.2	1506.9
1507.6	1506.2
1507.7	1506.1
1507.8	1506.5
1507.9	1506.6
Section 1508	**Section 1509**
1508.1 (New)	None
1508.2	1509.6
1508.3	1509.2
1508.3.1	1509.3
1508.4 (New)	None
1508.4.1	1509.1
1508.4.2	1509.5
1508.4.3	1509.4
1508.4.4	1509.8
1508.4.5	1509.9
1508.4.6	1509.11
1508.4.7	1509.7
1508.5	1509.10
Section 1509	**Section 1511**
1509.1	1511.1
1509.2	1511.2
1509.3	1511.3
1509.4 (New)	None
1509.4.1	1511.7
1509.4.2	1511.8
1509.4.2.1	1511.8.1
1509.4.3	1511.6
1509.5	1511.4
1509.6	1511.5
1509.6.1	1511.5.1
Section 1510	**Section 1510**
1510.1	1510.1
1510.2	1510.4
1510.3	1510.2
1510.4	1510.5
1510.5	1510.3

1504.2

Location of Spray-Finishing Operations

CHANGE TYPE. Modification

CHANGE SUMMARY. Section 1504.2 was revised to exempt spray-on automotive lining operations, such as those used in coating pickup truck beds, from requirements that govern other spray finishing operations. Other modifications to this section dealing with limited spraying areas and resin application areas are described in the previous Chapter 15 rewrite discussion. **(F178-04/05) and (F179-04/05)**

2006 CODE: ~~1504.1~~1504.2 **Location of Spray-Finishing Operations.** Spray-finishing operations conducted in buildings used for Group A, E, I or R occupancies shall be located in a spray room protected with an approved automatic sprinkler system installed in accordance with Section 903.3.1.1 and separated vertically and horizontally from other areas in accordance with the *International Building Code.* In other occupancies, spray-finishing operations shall be conducted in a spray room, spray booth, or ~~limited~~ spraying space approved for such use.

Exceptions:

~~**1508.1 General**~~

1. Automobile undercoating spray operations <u>and spray-on automotive lining operations</u> conducted in areas with approved natural or mechanical ventilation shall be exempt from the provisions of Section 1504 when approved and where utilizing Class IIIA or IIIB combustible liquids.

2. <u>In buildings other than Group A, E, I or R occupancies, approved limited spraying space in accordance with Section 1504.9.</u>

Requirements for spray application of automotive lining materials and resin materials used to manufacture reinforced plastics have been clarified.

3. Resin application areas used for manufacturing of reinforced plastics complying with Section 1509 shall not be required to be located in a spray room, spray booth or spraying space.

CHANGE SIGNIFICANCE. Automobile undercoating is no longer covered in a separate section (previously 1508) since the purpose of that section was limited to providing an exception to spray finishing regulations. The provisions are now more suitably located in Exception 1 to Section 1504.2. In addition, the provisions were expanded to encompass spray-on automotive lining operations.

Spraying of Class IIIA and Class IIIB liquids utilized as automotive linings, such as bed liners for pick-up trucks, is not believed to present a level of hazard that warrants protection using a spray booth, spray room, or spraying space. Instead the fire hazard is regarded as being relatively equivalent to undercoating operations. To reflect this in the code, a reference to spray-on automotive lining operations was added to Exception 1.

Some fire officials have reportedly seen results from testing conducted by the automotive lining industry demonstrating that spray-applied lining operations do not present a fire hazard. However, no such information was submitted for review by the ICC membership when this code change was considered, and the actual fire risk associated with these operations remains unknown.

It is overly simplistic to conclude that automotive lining operations do not present a fire hazard simply because they involve Class IIIA and IIIB liquids. Although it is true that spills of Class IIIA and IIIB liquids are not readily ignitable, the fire hazard increases when these liquids are sprayed because of an increase in the surface area to mass ratio. Because flammable and combustible liquid classifications are not a reliable hazard indicator for sprayed liquids (test methods associated with liquid classifications are based on liquid in a container, as opposed to sprayed liquids), there was no clear reason for the ICC membership to have approved this change without comprehensive technical substantiation. Nevertheless, the change was approved.

Exceptions 2 and 3 were added as part of the overall Chapter 15 rewrite and are discussed in the general discussion of Chapter 15 herein.

1504.7.6

Ventilation Termination Point

CHANGE TYPE. Deletion

CHANGE SUMMARY. Provisions regulating permissible discharge locations for environmental air ducts previously included in Exception 3 were deleted. **(F143-03/04)**

2006 CODE: ~~**1504.2.6**~~ **1504.7.6 Termination Point.** The termination point for exhaust ducts discharging to the atmosphere shall not be less than the following distances:

1. **and 2.** (No change to current text)

3. ~~Environmental air duct exhaust: 3 feet (914 mm) from the property line; 3 feet (914 mm) from openings into the building.~~

CHANGE SIGNIFICANCE. Exception 3, which dealt with environmental air exhaust, was deleted because it was inconsistent with the scope of Section 1504, and Chapter 15 in general. In the 2006 edition of the IMC, a definition was added to clarify that environmental air is "air that is conveyed to or from occupied areas through ducts which are not part of the heating or air-conditioning system, such as ventilation for human usage, domestic kitchen range exhaust, bathroom exhaust and domestic clothes dryer exhaust." Clearly, this is not germane to flammable finishing operations, so Exception 3 was out of place in Chapter 15.

Although they no longer appear in Chapter 15, regulations governing environmental air exhaust were not altogether deleted. They can still be found in IMC Sections 401.4 and 501.2.1; although, they do not apply to ventilation systems serving flammable finishing operations.

The IFC no longer regulates environmental air exhaust termination points. Only termination points for exhaust ducts serving flammable finishing operations, such as the ones shown above, are regulated by Chapter 15.

CHANGE TYPE.   Modification

CHANGE SUMMARY.   Increased quantities of hazardous production materials classified as oxidizer liquids, toxic liquids, pyrophoric solids, and water-reactive solids are now permitted in semiconductor fabrication areas to accommodate processing of larger silicon wafers and to accommodate new process technologies. **(F146-03/04, F182-04/05)**

2006 CODE:

TABLE 1805.2.2   **Maximum Quantities of HPM at a Workstation[e]** (revised portions only)

HPM Classification	State	Maximum Quantity
Oxidizer	Gas	3 cylinders
	Liquid	Use-open system 12 gallons[c] Use-closed system 60 gallons[a, c] ~~12 gallons[a, b, c]~~
	Solid	20 pounds[b,c]
Pyrophoric	Liquid	0.5 gallon[d, g]
	Solid	See Table 1804.2.2.1
Toxic	Liquid	Use-open system 15 gallons[c] Use-closed system 60 gallons[a,c] ~~15 gallons[a, b, c]~~
	Solid	5 pounds[b,c]
Water-reactive Class 3 Liquid	Liquid	0.5 gallon[d, g]
	Solid	See Table 1804.2.2.1

d. Allowed only in workstations that are internally protected with an approved automatic fire-extinguishing or <u>fire protection</u> ~~suppression~~ system complying with Chapter 9 <u>and compatible with the reactivity of materials in use at the workstation.</u>

g. <u>A maximum quantity of 5.3 gallons shall be allowed at a workstation when conditions are in accordance with Section 1805.2.3.5.</u>

   (Portions of table not shown did not change)

1805.2.3.5 Pyrophoric Liquids and Class 3 Water-Reactive Liquids. <u>Pyrophoric liquids and Class 3 water-reactive liquids in containers greater than 0.5-gallon (2 L) but not exceeding 5.3-gallon (20 L) capacity shall be allowed at workstations when located inside cabinets and the following conditions are met:</u>

1. <u>Maximum amount per cabinet: The maximum amount per cabinet shall be limited to 5.3 gallons (20 L).</u>

2. <u>Cabinet construction: Cabinets shall be constructed in accordance with the following:</u>
   2.1. <u>Cabinets shall be constructed of not less than 0.097-inch (2.5 mm)(12 gauge) steel.</u>

Table 1805.2.2 and Section 1805.2.3.5 continues

Table 1805.2.2 and Section 1805.2.3.5

Maximum Quantities of Hazardous Production Materials (HPMs) at a Workstation

*Table 1805.2.2 and Section
1805.2.3.5 continued*

2.2. Cabinets shall be permitted to have self-closing limited access ports or noncombustible windows that provide access to equipment controls.
2.3. Cabinets shall be provided with self- or manual-closing doors. Manual-closing doors shall be equipped with a door switch that will initiate local audible and visual alarms when the door is in the open position.

3. Cabinet exhaust ventilation system: An exhaust ventilation system shall be provided for cabinets and shall comply with the following:
   3.1. The system shall be designed to operate at a negative pressure in relation to the surrounding area.
   3.2. The system shall be equipped with a pressure monitor and a flow switch alarm monitored at the on-site emergency control station.

4. Cabinet spill containment: Spill containment shall be provided in each cabinet, with the spill containment capable of holding the contents of the aggregate amount of liquids in containers in each cabinet.

5. Valves: Valves in supply piping between the product containers in the cabinet and the workstation served by the containers shall fail in the closed position upon power failure, loss of exhaust ventilation and upon actuation of the fire control system.

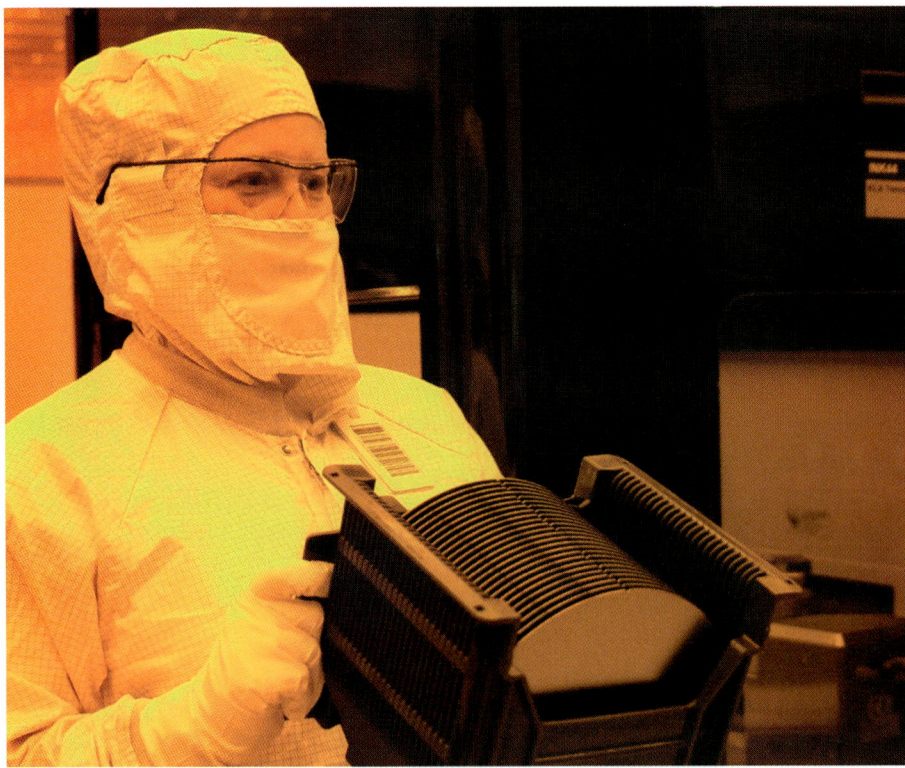

Semiconductor fabrication areas are now permitted to have larger quantities of certain hazardous production materials to accommodate new manufacturing technologies. (Photo courtesy of Spansion, Copyright © Spansion LLC.)

6. Fire detection system: Each cabinet shall be equipped with an automatic fire detection system complying with the following conditions:

> **6.1.** Automatic detection system: UV/IR, high sensitivity smoke detection (HSSD) or other approved detection systems shall be provided inside each cabinet.
>
> **6.2.** Automatic shutoff: Activation of the detection system shall automatically close the shutoff valves at the source on the liquid supply.
>
> **6.3.** Alarms and signals: Activation of the detection system shall initiate a local alarm within the fabrication area and transmit a signal to the emergency control station. The alarms and signals shall be both visual and audible.

CHANGE SIGNIFICANCE.   Changes made to Table 1805.2.2 and Section 1805.2.3.5 can be categorized into three groups: (1) increased quantity limits for oxidizer liquids and toxic liquids, (2) increased quantity limits for pyrophoric solids and Class 3 water-reactive solids, and (3) clarification of the fire protection requirements for workstations. These are discussed separately next.

1. **Increased quantity limits for oxidizer liquids and toxic liquids.** With the move to 300 mm wafer fabrication technology, the capacity of chemicals baths in semiconductor production tools must be increased to accommodate larger wafers. Prior workstation quantity limits for HPM oxidizer liquids and HPM toxic liquids date back more than 20 years, when wafer sizes were $\frac{1}{3}$ of those produced with current technology. Safety control and tool design technologies have improved dramatically since the original quantity limits were put in place, and with the quantity increases being limited to closed systems, it was considered reasonable to permit these increases, which were requested by the semiconductor manufacturing industry.

2. **Increased quantity limits for pyrophoric solids and Class 3 water-reactive solids.** As the semiconductor industry progresses to smaller production line widths and increased device densities, newer materials are demanded. Conducting and nonconducting structures, now fabricated at atomic levels, use a new class of materials known as metal-organics, which may possess both pyrophoric and water-reactive properties. Some common metal-organics used in the industry are trimethyl aluminum (TMA), trimethyl gallium (TMG), and dimethyl aluminum hydride (DMAH).

   For some production areas, the 0.5-gallon container limit established by prior code editions required a change-out frequency of once every 15 days with approximately 8–10 hours of tool downtime for each change-out. This percentage of downtime was very costly to the industry, and such frequent replacements of the bulk source increased the risk of an incident due to more frequent handling of materials. Industry made the case that it was less hazardous overall to permit an

Table 1805.2.2 and Section 1805.2.3.5 continues

Table 1805.2.2 and Section 1805.2.3.5 continued

increased, but reasonable, quantity of material in the fabrication area in order to reduce the risk of a handling-related incident. The alternative of installing bulk distribution systems with larger quantities centralized in a Group H-2 occupancy would require extensive piping runs, which come with their own set of risks, including potential leaks from piping or fittings and accidental damage.

Proponents of this change indicated that Factory Mutual has no loss history associated with metal-organic materials at any site, including sites that use large volumes of these materials. In addition, they indicated that the semiconductor industry has no reported loss history associated with the use of water-reactive/pyrophoric liquids.

To offset the risk of having increased quantities in fabrication areas, additional safety control requirements were added to Section 1805.2.3.5.

3. **Clarification of the fire protection requirements for workstations.** Footnote d has been revised by replacing "fire suppression system," an undefined term, with "fire protection system," a term defined in Section 902. This revision clarifies the intent of Footnote d, which is to permit the use of systems that will control, but not necessarily extinguish, a fire. This change is necessary with respect to fire protection systems for semiconductor manufacturing tools because of unique fire protection approaches sometimes used with this type of equipment.

CHANGE TYPE. Modification

CHANGE SUMMARY. Information required to be posted on warning signs at motor vehicle fuel-dispensing stations has been revised to improve clarity, to add a warning related to static discharge, and to enhance the warning related to filling of portable containers. **(F183-04/05)**

2006 CODE: 2205.6 Warning Signs. Warning signs shall be conspicuously posted within sight of each dispenser in the fuel-dispensing area and shall state the following:

1. ~~It is illegal and dangerous to fill unapproved containers with fuel.~~
2. ~~2.~~1. No smoking ~~is prohibited.~~
3. ~~3.~~2. ~~The engine shall be shut off during the refueling process.~~ Shut off motor.
4. ~~Portable containers shall not be filled while located inside the trunk, passenger compartment, or truck bed of a vehicle.~~
3. Discharge your static electricity before fueling by touching a metal surface away from the nozzle.
4. To prevent static charge, do not reenter your vehicle while gasoline is pumping.
5. If a fire starts, do not remove nozzle—back away immediately.
6. It is unlawful and dangerous to dispense gasoline into unapproved containers.
7. No filling of portable containers in or on a motor vehicle. Place container on ground before filling.

2205.6 continues

2205.6
Warning Signs

A fuel dispenser with the required warning sign.

Signs warning persons dispensing motor fuels of the hazards of static discharge and filling unapproved portable containers are required at motor fuel dispensers.

2205.6 continued

CHANGE SIGNIFICANCE. This revision was developed by a task group consisting of code officials, petroleum industry representatives, and other fire and safety professionals. The primary focus of the change is to add a warning to advise users of the need to control the discharge of static electricity while dispensing gasoline into vehicles and portable containers.

In the past several years, a number of fires involving a flash ignition at the dispenser nozzle have occurred as a result of static discharge. A common element identified in many of these fires was the re-entry of the dispenser operator into the vehicle during the filling process. It has been determined that re-entry into a vehicle, which involves sliding across the seating surface, can generate a static charge adequate to ignite gasoline vapors when the charge jumps from an individual's hand to a dispenser nozzle.

The severity of the static charge can change from very low to very high depending on a number of factors, such as relative humidity, tightness/looseness of clothing, and the clothing and seating materials. Anyone who has experienced a shock when touching a car door after exiting a vehicle, typically on a cold day with low relative humidity, can attest to the amount of energy discharged.

The new sign is also more straightforward with respect to warning the public about the need to place portable containers on the ground before filling to provide an electrical grounding path. Numerous fires have been documented involving portable containers being filled in passenger compartments, trunks, and truck beds because carpeting, bed liners, etc., prevent the dissipation of static electricity caused by fuel splashing inside of containers when they are being filled.

In making this change to the IFC, it was the goal of the task group that authored the proposal to create a standardized sign that would be utilized throughout the United States. However, although similar changes were made to NFPA 30A, the NFPA standard on motor vehicle fuel-dispensing stations, there are still slight differences between the signage requirements in the two documents. For example, the NFPA 30A warning sign addresses dispensing of gasoline by youths who are not licensed drivers, but the IFC sign does not include this information. Instead, the IFC assumes that individual states will prescribe suitable supplemental warnings to be added on the warning sign to address dispensing of fuel by minors.

CHANGE TYPE. Modification

CHANGE SUMMARY. Required separations between vaults containing aboveground tanks of flammable and combustible liquids and adjacent buildings, property lines, and fuel dispensers have been reduced to zero in recognition of the high level of fire resistance provided by vault enclosures. **(F150-03/04)**

2006 CODE:

Table 2206.2.3

Minimum Separation Requirements for Aboveground Tanks

TABLE 2206.2.3 Minimum Separation Requirements for Aboveground Tanks

Class of Liquid and Tank Type	Individual Tank Capacity (gallons)	Minimum Distance from Nearest Important Building on Same Property (feet)	Minimum Distance from Nearest Fuel Dispenser (feet)	Minimum Distance from Lot Line that Is or Can be Built Upon, Including the Opposite Side of a Public Way (feet)	Minimum Distance from Nearest Side of Any Public Way (feet)	Minimum Distance Between Tanks (feet)
Class I protected aboveground tanks ~~or tanks in vaults~~	Less than or equal to 6000	5	25[a]	15	5	3
	Greater than 600015	15	25[a]	25	15	3
Class II and III protected aboveground tanks ~~or tanks in vaults~~	Same as Class I	Same as Class I	Same as Class I	Same as Class I	Same as Class I	Same as Class I
Tanks in vaults	0–20,000	0[b]	0	0[b]	0	Separate compartment required for each tank
Other tanks	All	50	50	100	50	3

For SI: 1 foot = 304.8 mm, 1 gallon = 3.785 L.

a. At fleet vehicle motor fuel-dispensing facilities, no minimum separation distance is required.

b. Underground vaults shall be located such that they will not be subject to loading from nearby structures, or they shall be designed to accommodate applied loads from existing or future structures that can be built nearby.

CHANGE SIGNIFICANCE. The required separation distances for above grade vaults have been reduced on the basis of similarity of hazard between a tank in a vault and a tank in a building. Because vaults are always required to be entirely constructed of not less than 6-inch-thick reinforced concrete, a tank in a vault is essentially equivalent to a tank in an unoccupied, all-concrete, fire-resistive building.

Recognizing that a 6-inch-thick concrete wall is typically equated to a fire-resistance rating of 4 hours, the IBC permits a concrete building with an occupancy classified as Group H, such as a tank storage building, to be located at a property line. NFPA 30 contains a similar allowance for buildings with a blank wall having a fire-resistance rating of not less than 4 hours, so there is substantial precedence for permitting the "zero" separation distances now specified for vaults in Table 2206.2.3.

Table 2206.2.3 continues

Table 2206.2.3 continued

It should be noted that Table 2206.2.3 also addresses required separation distances for aboveground tanks installed in below grade vaults. For these installations, no separation is required between the vault enclosure and adjacent foundations, basement walls, or property lines, provided that the vault is capable of handling loads that may be applied by current or future buildings, as required by the new Footnote "b."

As a result of the changes made to Table 2206.2.3, required vault separation distances in the IFC are now consistent with NFPA 30, Flammable and Combustible Liquids Code, and NFPA 30A, Code for Motor Fuel Dispensing Facilities and Repair Garages, meaning that applicable regulations are now standardized among all current U.S. model codes that regulate vault installations.

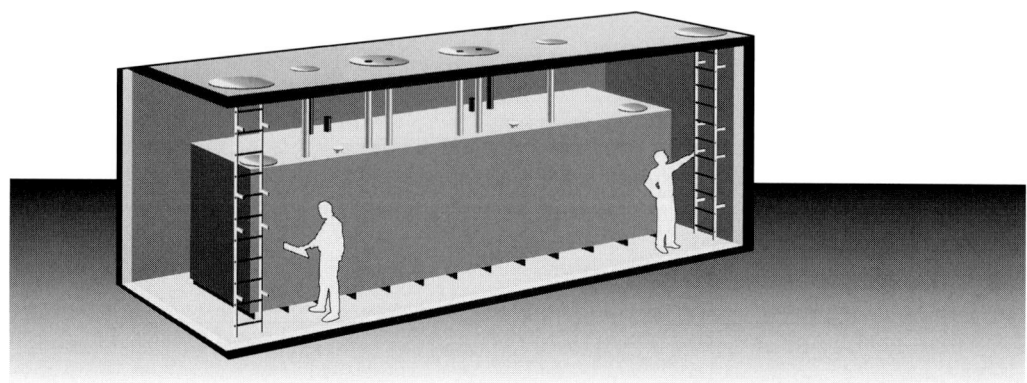

Cut-away view of an aboveground tank vault. A vaulted aboveground storage tank does not require separation from a building or property line. (Figure courtesy of Oldcastle Precast, Inc.)

2209.3.2 and 2211

Location of Dispensing Operations and Equipment for Hydrogen Motor Fuel

CHANGE TYPE.   Modification

CHANGE SUMMARY.   A number of changes have been made to Chapter 22 to facilitate adoption of hydrogen as a motor fuel. Changes to Section 2209.3.2 provide new alternatives for locating hydrogen generation, compression, storage, and dispensing operations. Changes to Section 2211 update provisions for repair garages to better accommodate hydrogen-fueled vehicles. **(F152-03/04, F154-03/04, F155-03/04, F157-03/04, F187-04/05, F188-04/05, F192-04/05)**

2006 CODE:

2209.3.2 Location of Dispensing Operations and Equipment. Generation, compression, storage and dispensing equipment shall be located ~~outdoors, above ground~~ in accordance with Sections 2209.3.2.1 through 2209.3.2.6.3.

~~**Exception:** 1.~~

2209.3.2.1 Outdoors. Generation, compression, storage or dispensing equipment shall be allowed ~~in buildings of Type I and II construction, as defined in the~~ *~~International Building Code,~~* ~~which are unenclosed for three-quarters or more of the perimeter and constructed in a manner that prevents the accumulation of hydrogen gas~~ outdoors in accordance with Section 2209.3.1.

~~**Exception:** 2.~~

2209.3.2.2 Weather Protection. Generation, compression, storage ~~and~~ or dispensing equipment shall be allowed ~~indoors~~ under

2209.3.2 and 2211 continues

A hydrogen fuel dispenser. (Photo courtesy of Honda Motor Corporation.)

The 2006 IFC contains requirements for the generation, compression, storage, and dispensing of hydrogen as a motor fuel. (Photo courtesy of Honda Motor Corporation.)

2209.3.2 and 2211 continued

weather protection in accordance with ~~Chapter 30 and as set forth in the~~ *~~International Building Code~~* ~~and~~ *~~International Fuel Gas Code~~* the requirements of Section 2704.13 and constructed in a manner that prevents the accumulation of hydrogen gas.

2209.3.2.3 Indoors. Generation, compression, storage and dispensing equipment shall be located in indoor rooms or areas constructed in accordance with the requirements of the *International Building Code,* the *International Fuel Gas Code,* and the *International Mechanical Code* and one of the following:

1. Inside a building in a hydrogen cutoff room designed and constructed in accordance with Section 420 of the *International Building Code.*

2. Inside a building not in a hydrogen cutoff room where the gaseous hydrogen system is listed and labeled for indoor installation and installed in accordance with the manufacturer's installation instructions.

3. Inside a building in a dedicated hydrogen fuel dispensing area having an aggregate hydrogen delivery capacity no greater than 12 standard cubic feet per minute (SCFM) and designed and constructed in accordance with Section 703.1 of the *International Fuel Gas Code.*

2209.3.2.3.1 Maintenance. Gaseous hydrogen systems and detection devices shall be maintained in accordance with the manufacturer's instructions.

2209.3.2.3.2 Smoking. Smoking shall be prohibited in hydrogen cutoff rooms. "No Smoking" signs shall be provided at all entrances to hydrogen cutoff rooms.

2209.3.2.3.3 Ignition Source Control. Open flames, flame-producing devices and other sources of ignition shall be controlled in accordance with Chapter 35.

2209.3.2.3.4 Housekeeping. Hydrogen cutoff rooms shall be kept free from combustible debris and storage.

2209.3.2.4 Gaseous Hydrogen Storage. Storage of gaseous hydrogen shall be in accordance with Chapters 30 and 35.

2209.3.2.5 Liquefied Hydrogen Storage. Storage of liquefied hydrogen shall be in accordance with Chapter 32.

2209.3.2.5.1 Location on Property. In addition to the requirements of Section 2203.1, above-ground liquefied hydrogen storage containers, compression and vaporization equipment serving motor fuel-dispensing operations shall be located 25 feet (7620 mm) from buildings having combustible exterior wall surfaces; buildings having noncombustible exterior wall surfaces that are not part of a 1-hour fire-resistance-rated assembly; wall openings; lot lines of property that could be built on; public streets; and parked vehicles.

2209.3.2.5.1.1 Barrier Wall Construction–Liquefied Hydrogen.

The outdoor separation distance shall be permitted to be reduced to 5 feet (1524 mm) where a 2-hour fire barrier interrupts the line of sight between equipment, other than dispensers, and the exposure within the radial distance as indicated by the tabular value. The height of the barrier shall be a minimum of 6 feet (1829 mm) but no less than 1.5 times the height of equipment, other than the cryogenic storage vessel, measured vertically. The length of the wall shall be no less than 1.5 times the maximum diameter or length of the tank. The 2-hour fire barrier shall not have more than two sides at approximately 90-degree (1.57 rad) directions, or three sides with connecting angles of approximately 135 degrees (2.36 rad). When fire barrier walls on three sides are used, piping and control systems serving stationary tanks shall be located at the open side of the enclosure created by the barrier walls.

2209.3.2.5.1.2 Location of Equipment.

Equipment shall be located from the enclosing walls at a distance not less than one tank diameter. When horizontal tanks are used the distance from any one enclosing wall shall be not less than one-half the length of the tank or a minimum of 5 feet (1524 mm).

2209.3.2.6 Canopy Tops.

Gaseous hydrogen compression and storage equipment located on top of motor fuel-dispensing facility canopies shall be in accordance with Sections 2209.3.2.6.1 through 2209.3.2.6.3, Chapters 30 and 35 and the *International Fuel Gas Code.*

2209.3.2.6.1 Construction.

Canopies shall be constructed in accordance with the motor fuel-dispensing facility canopy requirements of Section 406 of the *International Building Code.*

2209.3.2.6.2 Fire-Extinguishing Systems.

Fuel-dispensing areas under canopies shall be equipped throughout with an approved automatic sprinkler system in accordance with Section 903.3.1.1. The design of the sprinkler system shall not be less than that required for Extra Hazard Group 2 occupancies. Operation of the sprinkler system shall activate the emergency functions of Sections 2209.3.2.6.2.1 and 2209.3.2.6.2.2.

2209.3.2.6.2.1 Emergency Discharge.

Operation of the automatic sprinkler system shall activate an automatic emergency discharge system, which will discharge the hydrogen gas from the equipment on the canopy top through the vent pipe system.

2209.3.2.6.2.2 Emergency Shutdown Control.

Operation of the automatic sprinkler system shall activate the emergency shutdown control required by Section 2209.5.3.

2209.3.2 and 2211 continues

2209.3.2 and 2211 continued

2209.3.2.6.3 Signage. Approved signage having 2-inch (51 mm) block letters shall be affixed at approved locations on the exterior of the canopy structure stating: CANOPY TOP HYDROGEN STORAGE.

SECTION 2211. REPAIR GARAGES
(Sections 2211.1 through 2211.7.1, no changes.)

2211.7.1.1 Design. Indoor locations shall be ventilated utilizing air supply inlets and exhaust outlets arranged to provide uniform air movement to the extent practical. Inlets shall be uniformly arranged on exterior walls near floor level. Outlets shall be located at the high point of the room in exterior walls or the roof.

Ventilation shall be by a continuous mechanical ventilation system or by a mechanical ventilation system activated by a continuously monitoring natural gas detection system ~~where~~ or, for hydrogen, a continuously monitoring flammable gas detection system, each activating at a gas concentration of not more than 25 percent of the lower flammable limit (LFL) ~~is present~~. In ~~either~~ all cases, the system shall shut down the fueling system in the event of failure of the ventilation system.

The ventilation rate shall be at least 1 cubic foot per minute per 12 cubic feet $(0.00139 \text{ m}^3 \times \text{m}^3)$ of room volume.

(Sections 2211.7.1.2 through 2211.8.3.1 have no changes.)

CHANGE SIGNIFICANCE. In the past several years, a great deal of effort has gone into modifying the IFC to help mainstream the use of hydrogen gas as a motor fuel. The 2003 edition of the IFC was the first edition to include unique provisions for hydrogen in Chapter 22, and the 2006 edition has enhanced and expanded these provisions. Proponents of hydrogen fuel make the case that, in general, hydrogen fuel is no more dangerous than gasoline, propane, methane, or other fuels. In an outdoor environment, gaseous hydrogen is arguably safer because it tends to disperse more rapidly than methane, propane, or gasoline vapor.

Many of the 2006 edition changes relate to needed infrastructure to facilitate generation, storage, and dispensing of hydrogen at motor vehicle fueling stations. Other changes help mitigate hazards associated with having hydrogen-fueled vehicles in repair garages.

With respect to generation, storage, and dispensing of hydrogen, Section 2209.3.2 has been modified to expand and better address permissible locations for these operations. Generation, compression, storage, and dispensing equipment are now permitted to be:

1. Outdoors (2209.3.2.1), and to a limited degree, outdoors on top of a canopy covering dispensers (2209.3.2.6);

2. Outdoors, under a weather-protection canopy (2209.3.2.2);

3. Indoors (2209.3.2.3); or

4. In some cases, in an aboveground vault or underground, either direct-buried or in a vault per Chapters 30 and 32.

Section 2209.3.2.4 prescribes additional requirements for gaseous hydrogen via reference to IFC Chapters 30 and 35. Except where conflicting regulations are included in Chapter 22, Chapters 30 and 35 apply to gaseous hydrogen. Because Chapter 30 permits the use of aboveground or underground gas vaults, gas vaults are an option for gaseous hydrogen storage, even though they are not specifically mentioned in Chapter 22.

Section 2209.3.2.5 prescribes additional requirements for liquefied hydrogen via reference to IFC Chapter 32. Except where conflicting regulations are specified in Chapter 22 for liquefied hydrogen, Chapter 32 applies. Because Chapter 32 permits the use of direct-buried underground tanks, underground tanks are an option for liquefied hydrogen storage even though they are not specifically mentioned in Chapter 22.

Generally speaking, the more enclosed a hydrogen generation, storage, or dispensing operation becomes, the greater the hazard, because enclosures capable of trapping gas increase the likelihood of accumulating gas in the flammable range. By placing operations in areas where gas accumulation is inhibited, risk is reduced. Accordingly, the code tends to be less restrictive for outdoor operations than for indoor operations, with one exception: storage on top of canopies.

Locating hydrogen equipment on top of a canopy covering dispensing operations is somewhat controversial. From the industry's perspective, the approach is appealing because it makes cost-effective use of real estate. There is no need to dedicate a portion of the property to hydrogen equipment, either inside of a building or outside, and unlike buried equipment, canopy-top equipment is readily accessible for inspection and repair. Conversely, the perspective of some code officials is that a fire beneath a canopy may place equipment above at risk.

For now, the code has adopted a compromise that permits hydrogen equipment on top of canopies but requires a number of special controls to offset safety concerns. To address these special controls for canopy-top storage, Section 2209.3.2.6 has been added. Included among the controls are a requirement to provide fire sprinklers to protect areas beneath the canopy and a requirement to automatically release stored hydrogen upon activation of the sprinkler system below.

Although not specifically stated by the code, the intent is to limit canopy-top storage to gaseous hydrogen, as opposed to liquefied hydrogen. Section 2209.3.2.6 is specific to gaseous hydrogen equipment, and by not mentioning liquefied hydrogen in this section, there is no intent to permit it.

In repair garages, which are governed by Section 2211, the use of hydrogen as a motor fuel necessitates special treatment. Special ventilation controls are needed because hydrogen is a lighter-than-air gas that is often nonodorized, meaning that leaks may go undetected until the concentration of leaked gas reaches the flammable range. Traditional repair garage designs, which place ventilation exhaust intakes at floor level and sources of ignition well above the floor based on heavier-than-air fuel vapors, are inappropriate to mitigate hazards associated with lighter-than-air fuels.

2209.3.2 and 2211 continues

2209.3.2 and 2211 continued   The change to Section 2211 in the 2006 edition recognizes the permissible use of a flammable-gas detection system to activate ventilation in lieu of providing continuous ventilation for hydrogen. As long as the ventilation system is activated well in advance of accumulating hydrogen in the flammable range, there is no need to operate ventilation continuously, which is very inefficient from an energy conservation perspective in some climates.

2308.2.1
Plastic Pallets and Shelves in Rack Storage

CHANGE TYPE. Modification

CHANGE SUMMARY. Section 2308.2.1 has been modified to permit storage on plastic pallets to be protected as required for storage on wood pallets when the plastic pallets are listed in accordance with UL 2335. **(F194-04/05)**

2006 CODE: 2308.2.1 Plastic Pallets and Shelves. Storage on plastic pallets or plastic shelves shall be protected by approved specially engineered fire protection systems.

> **Exception:** Plastic pallets listed and labeled in accordance with UL 2335 shall be treated as wood pallets for determining required sprinkler protection.

CHANGE SIGNIFICANCE. This revision provides a reference to recognize the reduced fire hazard associated with some plastic pallets. The new exception recognizes UL 2335, *Fire Tests of Storage Pallets,* as a basis for testing and listing of reduced-hazard pallets. Chapter 45 has also been modified to add a reference to adopt UL 2335. At the time that this code change was considered by the ICC membership, the code change proponent indicated that five companies had pallets listed to UL 2335.

2308.2.1 continues

Goods stored on plastic pallets that are listed and labeled as meeting UL 2335 are allowed to be treated as goods stored on wood pallets, for rack storage of commodities. (Photo courtesy of Buckhorn Incorporated).

2308.2.1 continued

There is precedence for recognizing these types of pallets in the NFPA 13–2002 edition, Section 12.1.9.2.1 (6), which permits non-wood pallets to be protected in the same manner permitted for wood when the nonwood pallets present a fire hazard not exceeding that of wood. However, this provision applies only to idle pallets, as opposed to those that are being used with stored commodities. In-service nonwood pallets are regulated by NFPA 13–2002, Section 5.6.2, which prescribes specific requirements for protecting commodities stored on these pallets. Note that NFPA 13 is moving to eliminate the use of the term "plastic pallet" in favor of the term "nonwood pallet."

Overall, the consequence of the new exception to Section 2308.2.1 is that, for racked storage, which is the topic governed by Section 2308, sprinkler protection is permitted to be designed on the basis of wood pallets as opposed to plastic when the pallets are listed per UL 2335. Although this is not consistent with provisions in the 2002 edition of NFPA 13, which ordinarily governs sprinkler design based on the adoption reference in IFC Chapter 45, representatives of Underwriters Laboratories who served as the proponent of this change indicated that the IFC approach is appropriate because UL requires comparable fire performance of nonwood versus wood pallets loaded with commodity as part of the UL 2335 listing evaluation. Because the IFC prevails when there is a conflict between the IFC and referenced standards, Section 2308.2.1 takes precedence over conflicting provisions in NFPA 13.

It must be emphasized that this change has no bearing on storage arrangements other than rack storage because the change falls under Section 2308.

PART 5

Hazardous Materials
Chapters 27 Through 44

- **Chapter 27**   Hazardous Materials
- **Chapter 28**   Aerosols
- **Chapter 29**   Combustible Fibers
- **Chapter 30**   Compressed Gases
- **Chapter 31**   Corrosive Materials No changes addressed
- **Chapter 32**   Cryogenic Fluids No changes addressed
- **Chapter 33**   Explosives and Fireworks
- **Chapter 34**   Flammable and Combustible Liquids
- **Chapter 35**   Flammable Gases
- **Chapter 36**   Flammable Solids No changes addressed
- **Chapter 37**   Highly Toxic and Toxic Materials
- **Chapter 38**   Liquefied Petroleum Gases
- **Chapter 39**   Organic Peroxides No changes addressed
- **Chapter 40**   Oxidizers No changes addressed
- **Chapter 41**   Pyrophoric Materials No changes addressed
- **Chapter 42**   Pyroxylin (Cellulose Nitrate) Plastics No changes addressed
- **Chapter 43**   Unstable (Reactive) Materials No changes addressed
- **Chapter 44**   Water-Reactive Solids and Liquids No changes addressed

Chapters 27 through 44 contain requirements for the safe storage, use, and handling of hazardous materials. The Chemical Abstract Service, a part of the American Chemical Council, has identified over ½ million commercial chemicals and mixtures. The *International Fire Code* does not regulate all of these chemicals. Instead the *International Fire Code* focuses on chemicals and chemical mixtures that are classified as certain physical or health hazards. Physical hazard materials can detonate, deflagrate, accelerate burning, or burn. Health hazard materials are chemicals or mixtures of chemicals that can incapacitate, injure, or cause death after only a single exposure. ■

SECTION 2701.1, EXCEPTION 10

Hazardous Materials—Scope

TABLE 2703.1.1(1) AND SECTION 2702.1

Maximum Allowable Quantities per Control Area for Explosives

TABLE 2703.1.1(1)

Maximum Allowable Quantities per Control Area for Combustible Fibers

TABLE 2703.1.1(1)

Maximum Allowable Quantities per Control Area for Fuel in Fuel Tanks and Piping Systems

2703.2.9

Testing of Hazardous Materials Equipment, Devices, and Systems

TABLE 2703.8.3.2

Design and Number of Control Areas

2703.8.3.4

Fire-Resistance Rating Requirements for Control Areas

CHAPTER 28

Aerosols

SECTION 2902.1

Definitions for Combustible Fibers

3003.3

Pressure Relief for Compressed Gas Containers, Cylinders, and Tanks

3003.16

Vaults for Compressed Gases

CHAPTER 33

Definition of Explosives

CHAPTER 33

Separation Distances for Explosives

CHAPTER 33

Separation Distances for Explosives—Day Boxes and Operating Buildings

3308.2

Permit Application for Fireworks

3308.9

Post–Fireworks Display Inspection

3402.1

Definition of Liquid Storage Warehouse

3404.2.7.5.8 AND 3406.4.6

Overfill Prevention

3404.3.1

Design, Construction, and Capacity of Containers and Portable Tanks

3404.3.2.3

Number of Storage Cabinets

3404.3.5.1

Basement Storage

TABLE 3404.3.6.3(2)

Storage of Unsaturated Polyester Resin

TABLE 3404.3.6.3(3)

Quantity Limits for Liquid Storage Warehouses

3405.3.8.4

Weather Protection

3405.5

Alcohol-Based Hand Rubs Classified as Class I or II Liquids

3504.2.1

Distance Limit to Exposures for Flammable Gases

3702.1, 3704.2.2.10 AND 1803.13

Physiological Warning Thresholds

3704.2.2.7

Treatment Systems for Toxic Gases

3806.2

Overfill Prevention for LP-Gas Containers

3809.12

LP-Gas Storage Outside of Buildings

Section 2701.1, Exception 10

Hazardous Materials— Scope

Alcohol-based hand rubs in wall-mounted dispensers are now exempt from the requirements of Chapter 27.

CHANGE TYPE.   Addition

CHANGE SUMMARY.   A new exception has been added to permit certain installations of alcohol-based hand rub dispensers to be exempted from the requirements of Chapter 27. **(F233-04/05)**

2006 CODE: 2701.1 Scope.   Prevention, control, and mitigation of dangerous conditions related to storage, dispensing, use, and handling of hazardous materials shall be in accordance with this chapter.

This chapter shall apply to all hazardous materials, including those materials regulated elsewhere in this code, except that when specific requirements are provided in other chapters, those specific requirements shall apply in accordance with the applicable chapter. Where a material has multiple hazards, all hazards shall be addressed.

Exceptions:

(No changes to 1–9.)

10. The use of wall-mounted dispensers containing alcohol-based hand rubs classified as Class I or II liquids when in accordance with Section 3405.5.

CHANGE SIGNIFICANCE.   The new Exception 10 provides an allowance to follow new provisions governing alcohol-based hand rub (ABHR) dispensers in Section 3405.5 in lieu of Chapter 27. It is the intent of this change to exempt the contents of liquids in mounted ABHR dispensers from the maximum allowable quantity limits for control areas prescribed in Chapter 27. However, ABHR that is not in mounted dispensers must still comply with Chapters 27 and 34.

Section 3405.5 was added in the 2006 edition to provide special regulations for ABHR dispensers, which have become an important part of efforts to control transmission of infectious diseases. These new regulations supersede general requirements that ordinarily govern flammable and combustible liquids. For additional information, see the discussion of Section 3405.5 herein.

CHANGE TYPE.   Modification

CHANGE SUMMARY.   Day boxes, typically used to contain explosives, are now recognized as an enclosure qualifying for increases in maximum allowable hazardous materials quantities in control areas. **(F163-03/04)**

2006 CODE:

TABLE 2703.1.1(1)   **Maximum Allowable Quantity per Control Area of Hazardous Materials Posing a Physical Hazard**[a, j, m, n, p]

(revised portion related to the new Footnote "e" only)

e. Maximum allowable quantities shall be increased 100 percent when stored in approved storage cabinets, <u>day boxes,</u> gas cabinets, exhausted enclosures, or safety cans. Where Note d also applies, the increase for both notes shall be applied accumulatively.

2702.1 <u>DAY BOX.</u>   <u>A portable magazine designed to hold explosive materials constructed in accordance with the requirements for a Type 3 magazine as defined and classified in Chapter 33.</u>

CHANGE SIGNIFICANCE.   The revised Footnote "e" permits quantity increases previously associated with storage cabinets, gas cabinets, exhausted enclosures, and safety cans to also apply to explosives in day boxes. A new definition of "day box" has been added to Section 2702.1 to correlate with this change.

Table 2703.1.1(1) and Section 2702.1 continues

Table 2703.1.1(1) and Section 2702.1

Maximum Allowable Quantities per Control Area for Explosives

Day boxes, which are portable explosive magazines, can be used to increase the maximum allowable quantity of hazardous materials in control areas. (Photo courtesy of Armag Corporation, Bardstown, KY.)

Table 2703.1.1(1) and Section 2702.1 continued

Section 3302.1 defines a Type 3 magazine as a portable, fire-resistant, theft-resistant, and weather-resistant "day box" or portable structure constructed in accordance with NFPA 495, NFPA 1124, or DOTy 27 CFR requirements. The term "day box" is used by the industry to describe magazines of this type. NFPA 495, Bureau of Alcohol, Tobacco and Firearms (BATF) requirements in 27 CFR and IFC Chapter 33 require Type 3 magazines to be fire-resistant, weather-resistant, and theft-resistant. They are to be constructed of not less than 12-gauge (0.1046) steel, lined with at least either ½-inch plywood or ½-inch Masonite type hardboard. Doors must overlap the openings by at least 1 inch. Hinges and hasps are required to be attached by welding, riveting, or bolting with nuts on the inside (theft-resistant), and a padlock is required.

Given that Table 2703.1.1(1) permits quantity increases for hazardous materials storage cabinets, gas cabinets, and exhausted enclosures, it is appropriate that Type 3 magazines be given the same level of recognition since they provide a comparable level of enclosure. A comparison of the basic required features is as follows:

Feature	Hazardous Materials Storage Cabinet Section 2703.8.7	Type 3 Magazine or Day Box (Indoors) Section 3304.2 and NFPA 495
Fire-resistant	18-gauge (1.2-mm), double-walled steel	12-gauge (2.5-mm), steel-lined, with ½-inch plywood or Masonite. Also Section 3304.5.1.2, Item 1
Weather-resistant	Not required	Required
Theft-resistant	Doors to be self-latching	Specified case-hardened, shackle padlock required. Hinges, hasps to be welded, riveted, or bolted from the inside. Also Section 3304.5.1.2, Item 1
Bottom of cabinet liquid-tight	2–inch-high liquid tight sill	Not required. Typical materials in solid form
Joints and hinges/hasps	Joints to be riveted or welded and tight-fitting	Hinges, hasps to be welded, riveted, or bolted from the inside
Doors	Well-fitted and self-closing	Doors to overlap opening by not less than 1 inch
Unattended Storage	Permitted	Not permitted. Used for operations only, with material to be removed to Type 1 or 2 magazines for unattended storage
Exterior paint	None required; however, the cabinet is required to be coated or otherwise nonreactive	Red; Section 3304.5.1.2, Item 2
Signage	Red letters on a contrasting background "Hazardous—Keep Fire Away"	White lettering: "Explosives—Keep Fire Away" Section 3304.5.1.2, Item 4
Dimensions	Least dimensions not specified	Least horizontal dimension shall not exceed the clear width of the entrance door; Section 3304.5.1.2, Item 5
Nonsparking materials	Not required	No exposed ferrous metal on the interior; Section 3304.6.4
Wheels, casters, or rollers	Not required	Indoor magazines to be fitted with the means to facilitate removal from the building in an emergency; Section 3304.5.1.2 3

CHANGE TYPE. Modification

CHANGE SUMMARY. Quantity limits for combustible fibers have been revised to exclude densely packed baled cotton. Fire tests have demonstrated that the burning characteristics of baled cotton are not severe enough to warrant regulating this material in the same manner as other combustible fibers. **(F204-04/05)**

2006 CODE:

TABLE 2703.1.1(1) **Maximum Allowable Quantity per Control Area of Hazardous Materials Posing a Physical Hazard**[a, j, m, n, p]

(revised portions related to combustible fibers only)	
Material	Class
Combustible fibers	Loose
	Baled[o]

o. Densely-packed baled cotton that complies with the packing requirements of ISO 8115 shall not be included in this material class.

CHANGE SIGNIFICANCE. A new Footnote "o" has been added to Table 2703.1.1(1) to exempt densely packed baled cotton from the quantity limits that otherwise apply to combustible fibers. This has the consequence of exempting densely packed baled cotton from ever triggering a Group H-3 occupancy classification, thereby allowing the material to be stored in Group S-1 occupancies, regardless of the quantity stored. The change was one of several revisions designed to discontinue treatment of densely packed baled cotton as a hazardous material, all of which were justified on the basis of fire tests. These tests demonstrated reduced burning characteristics of baled cotton as compared to other combustible fibers.

Table 2703.1.1(1) continues

Table 2703.1.1(1)

Maximum Allowable Quantities per Control Area for Combustible Fibers

Densely packed baled cotton, such as the baled cotton in this warehouse, is no longer regulated as a combustible fiber because of its limited burning characteristics.

Example of a cotton storage warehouse.

Table 2703.1.1(1) continued

Contrary to some historic anecdotal information regarding the combustibility characteristics of baled cotton, flammability research conducted by the industry and presented to the ICC membership showed that densely packed baled cotton (meeting the size and weight requirements of ISO 8115) does not warrant regulation as a hazardous material. According to the code change proponent, similar allowances have been made by the U.S. Department of Transportation, the U.S. Coast Guard, the International Maritime Organization (IMO), and NFPA.

In support of this revision, the code change proponent indicated the following:

1. Standard cotton fiber "passed" the Department of Transportation spontaneous combustion test, i.e., the cotton did not exceed the oven temperature and was not classified as self-heating.

2. Densely packed baled cotton does not support sustained smoldering propagation. Testing with an electric heater placed within a densely-packed bale was unable to cause sustained smoldering propagation due to the lack of oxygen inside the bale.

3. Densely packed baled cotton exposed to ignition from a cigarette and a match showed no propagating combustion with either.

4. Densely packed baled cotton exposed to ignition from the gas burner source specified in ASTM E 1590 (also known as California Technical Bulletin 129 [CTB 129]) passed all applicable criteria in the CTB 129 test, including mass loss, heat release rate, and total heat release during the first 10 minutes of the test. The CTB 129 test involves combustion of 12 liters per minute of propane gas for 3 minutes.

Table 2703.1.1(1)

Maximum Allowable Quantities per Control Area for Fuel in Fuel Tanks and Piping Systems

CHANGE TYPE.   Modification

CHANGE SUMMARY.   A new Footnote "p" has been added to clarify when it is permissible to exclude flammable and combustible liquids and flammable gases used as fuels from quantity calculations used to determine occupancy classification. **(F249-04/05)**

2006 CODE:

TABLE 2703.1.1(1)   Maximum Allowable Quantity per Control Area Of Hazardous Materials Posing a Physical Hazard[a, j, m, n, p]

(revised portion related to the new Footnote "p" only)

p. The following shall not be included in determining the maximum allowable quantities:

1. Liquid or gaseous fuel in fuel tanks on vehicles.
2. Liquid or gaseous fuel in fuel tanks on motorized equipment operated in accordance with this code.
3. Gaseous fuels in piping systems and fixed appliances regulated by the *International Fuel Gas Code.*
4 Liquid fuels in piping systems and fixed appliances, regulated by the *International Mechanical Code.*

Table 2703.1.1(1) continues

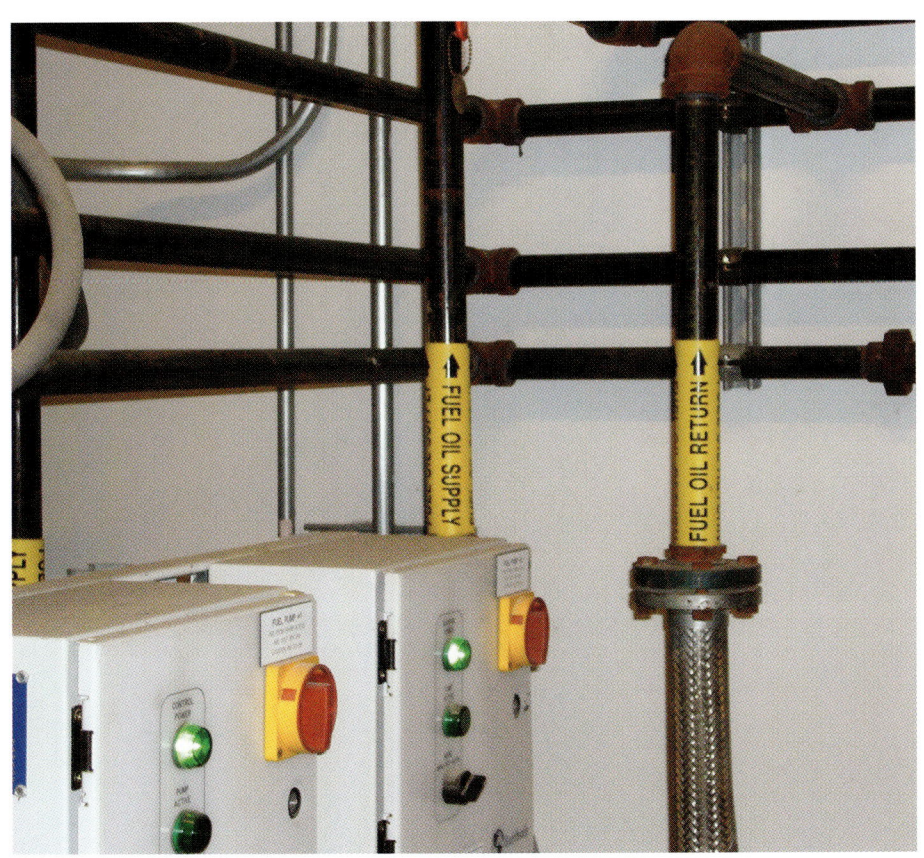

Liquids and gases used as fuels and located in piping systems and fixed appliances are among those that are no longer required to be included in calculations of the maximum allowable quantity in a control area.

Table 2703.1.1(1) continued

CHANGE SIGNIFICANCE. The new Footnote "p" provides additional emphasis on the intent of the code to exclude some flammable and combustible liquids and flammable gases used as fuels from quantity calculations used in determining occupancy classification. Specifically, fuels contained in fuel tanks of vehicles or motorized equipment and fuels contained in closed piping systems supplying fixed appliances regulated by the *International Fuel Gas Code* or the *International Mechanical Code* do not contribute to maximum allowable quantity (MAQ) calculations when determining whether a Group H occupancy is required.

Although this may appear to be a significant departure from the approach taken in previous code editions, that is not the case. A similar exclusion previously appeared in IBC Section 307.9, Exception 5, but because that exception was not referenced by the MAQ tables in IBC Section 307 and IFC Section 2703, it was often overlooked. From an intent standpoint, the code never intended that the quantity of fuel in vehicle fuel tanks (such as a gas tank on a car or forklift) or in closed piping systems (such as natural gas piping or piping feeding an emergency generator) be considered in MAQ calculations, so the new Footnote "p" is truly just a clarification of code intent.

One additional change has been made to the 2006 edition that further clarifies the code with respect to flammable and combustible liquids and flammable gases in piping systems. Exception 5 to the Group H occupancy definition in Section 202 has been revised to make it clear that flammable and combustible liquids and flammable gases in storage tanks and other portions of closed systems cannot be excluded from hazardous materials quantity calculations. The revised Exception 5 only permits exclusion of flammable and combustible liquids and flammable gases in closed piping systems. Additional information on this revision is provided in the discussion of Section 202 herein.

Fuel in vehicle fuel tanks is not required to be included in calculations of the maximum allowable quantity in a control area.

CHANGE TYPE. Addition

CHANGE SUMMARY. A new section has been added to establish minimum requirements for testing of certain hazardous materials equipment, devices, and systems. **(F206-04/05)**

2006 CODE:

2703.2.9.Testing. The equipment, devices and systems listed in Section 2703.2.9.1 shall be tested at one of the intervals listed in Section 2703.2.9.2. Written records of the tests conducted or maintenance performed shall be maintained in accordance with the provisions of Section 107.2.1.

Exceptions:
1. Testing shall not be required where approved written documentation is provided stating that testing will damage the equipment, device, or system and the equipment, device, or system is maintained as specified by the manufacturer.
2. Testing shall not be required for equipment, devices, and systems that fail in a fail-safe manner.
3. Testing shall not be required for equipment, devices, and systems that self-diagnose and report trouble. Records of the self-diagnosis and trouble reporting shall be made available to the authority having jurisdiction.

2703.2.9 continues

2703.2.9

Testing of Hazardous Materials Equipment, Devices, and Systems

Hazardous materials safety controls, such as this gas detector, are now subject to periodic inspections.

2703.2.9 continued

4. Testing shall not be required if system activation occurs during the required test cycle for the components activated during the test cycle.

5. Approved maintenance in accordance with Section 2703.2.6 that is performed not less than annually or in accordance with an approved schedule shall be allowed to meet the testing requirements set forth in Sections 2703.2.9.1 and 2703.2.9.2.

2703.2.9.1 Equipment, Devices, and Systems Requiring Testing. The following equipment, systems, and devices shall be tested in accordance with Sections 2703.2.9 and 2703.2.9.2.

1. Gas detection systems, alarms, and automatic emergency shutoff valves required by Section 3704.2.2.10 for highly toxic and toxic gases.

2. Limit controls systems for liquid-level, temperature, and pressure required by Sections 2703.2.7, 2704.8 and 2705.1.4.

3. Emergency alarm systems and supervision required by Sections 2704.9 and 2705.4.4.

4. Monitoring and supervisory systems required by Sections 2704.10 and 2705.1.6.

5. Manually activated shut down controls required by Section 4103.1.1.1 for compressed gas systems conveying pyrophoric gases.

2703.2.9.2 Testing Frequency. The equipment, systems and devices listed in Section 2703.2.9.1 shall be tested at one of the frequencies listed below.

1. Not less than annually;

2. In accordance with approved manufacturer's requirements;

3. In accordance with approved recognized industry standards; or

4. In accordance with an approved schedule.

CHANGE SIGNIFICANCE. The new Section 2703.2.9 provides necessary provisions for testing to ensure that the equipment, devices, and systems deemed critical to hazardous materials safety will perform as intended in an emergency. In addition, written records are required to document testing and maintenance.

Note that the code is very specific with regard to which equipment, devices, and systems must be tested, as set forth in Section 2703.2.9.1. Equipment, devices, and systems not listed in this section are not subject to mandatory testing under Section 2703.2.9, although testing may still be required to comply with provisions in reference standards or Federal laws, such as OSHA process safety management.

Even though equipment or a particular device or system may be listed as requiring testing in Section 2703.2.9.1, Section 2703.2.9 has a number of exceptions that overrule the provisions, including cases where testing might damage the unit to be tested, such as a nonresetting fixed-temperature heat detector used as a high-temperature limit control. For such equipment, testing is not required provided that the equipment is maintained as required by the manufacturer. In addition, equipment, devices, and systems that fail in the "safe" position, such as a normally closed automatic closing valve provided to stop flow in the event of an emergency, are not required to be tested. Nevertheless, it may be advisable to test valves of this type in some cases, regardless of the code's exception, because such valves may jam in the open position after an extended period of time without operating.

Exceptions are also permitted for equipment, devices, and systems that (1) are self-diagnosing, (2) have been activated by an event, or (3) receive maintenance at least annually in accordance with an approved maintenance schedule.

Provisions included in the new section were jointly developed by the Semiconductor Industry Association, the ICC/IAFC Western-Canadian Code Action Committee, and the Washington State Association of Fire Marshals, and they represent a multiyear effort in determining reasonable requirements with respect to what types of equipment, devices, and systems must be routinely tested and what the necessary testing frequency must be.

Table 2703.8.3.2

Design and Number of Control Areas

CHANGE TYPE. Modification

CHANGE SUMMARY. The limitation of two control areas per floor for Group M occupancies and certain Group S occupancies has been deleted. **(F165-03/04)**

2006 CODE:

TABLE 2703.8.3.2 Design and Number of Control Areas

(portions of table related to the deletion of Footnote "b")

a. Percentages shall be of the maximum allowable quantity per control area shown in Tables 2703.1.1(1) and 2703.1.1(2), with all increases allowed in the footnotes to those tables.

b. ~~There shall be a maximum of two control areas per floor in Group M occupancies and in buildings or portions of buildings having Group S occupancies with storage conditions and quantities in accordance with Section 2703.11.~~

~~c.~~ b. Fire barriers shall include walls and floors as necessary to provide separation from other portions of the building.

CHANGE SIGNIFICANCE. The limitation of two control areas per floor for Group M occupancies and certain Group S occupancies has been deleted. Control areas in these occupancies now follow the general rule applicable to other occupancies, which is four control areas on the first story and fewer above or below.

There was never a technical justification to uniquely limit the number of retail display and storage control areas, and the two control area limit had a significant impact on strip malls where multiple ten-

Group M and S occupancies are no longer limited to two control areas per floor.

ants, such as paint stores, drug stores, hardware stores, automotive stores, and other similar occupancies might have limited quantities of hazardous materials. From both compliance and enforcement perspectives, distributing maximum allowable quantities for control areas among multiple tenants proved impractical, and by increasing the permissible number of control areas from two to four, the problem is largely alleviated. With a four-control-area limit, several tenants can each have their own control area.

From a historic perspective, the two control area limit for mercantile occupancies dates back to the initial development of modern hazardous materials provisions that took place in the Uniform code arena in 1988. At the time, some believed that the quantity limits for mercantile occupancies should be more restrictive because of the potential for exposing customers during an incident.

However, in 1989, it was determined that the newly adopted quantity limits were so stringent that they technically forced many mercantile occupancies into the Group H occupancy category, even though a change of this nature had never been justified. To correct this problem, the Uniform codes were revised by significantly increasing permissible hazardous materials quantities in mercantile occupancies without triggering Group H classification.

It is important to note that the limitation of two control areas in prior editions of the IFC was not linked to the larger quantity limits permitted for these occupancies. On the contrary, the larger quantity limits were put in place after the two control area limit, and the two restrictions were based on entirely separate concerns.

2703.8.3.4

Fire-Resistance Rating Requirements for Control Areas

CHANGE TYPE. Modification

CHANGE SUMMARY. The required fire-resistance rating for floors supporting control areas has been reduced from 2 hours to 1 hour for buildings not exceeding three stories in height that are equipped with a fire sprinkler system and are of Type IIA, IIIA, or VA construction. **(F210-04/05)**

2006 CODE: 2703.8.3.3 Separation 2703.8.3.4 Fire-Resistance Rating Requirements. The required fire-resistance rating for fire barriers assemblies shall be in accordance with Table 2703.8.3.2. The floor construction of the control area and the construction supporting the floor of the control area shall have a minimum 2-hour fire-resistance rating.

> **Exception:** The floor construction of the control area and the construction supporting the floor of the control area is allowed to be 1-hour fire resistance rated in buildings of Type IIA, IIIA, and VA construction provided that both of the following conditions exist:
>
> 1. The building is equipped throughout with an automatic sprinkler system in accordance with Section 903.3.1.1, and
> 2. The building is three stories or less in height.

CHANGE SIGNIFICANCE. Some of the legacy codes did not require a 2-hour fire-resistance rating for floor assemblies supporting control areas, and the 2-hour requirement in the International Codes has had a particularly significant impact on low-height buildings, particularly existing buildings, which are often constructed using nonrated or 1-hour

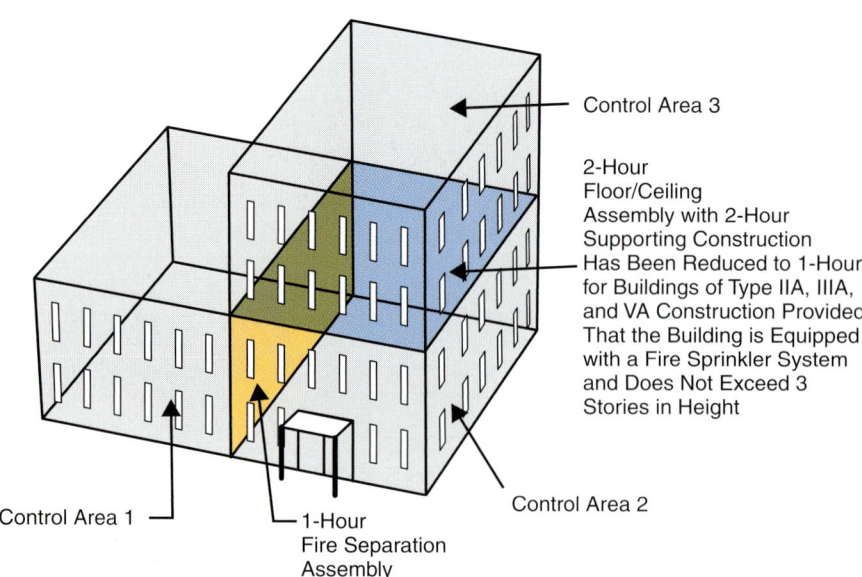

Control Area 3

2-Hour Floor/Ceiling Assembly with 2-Hour Supporting Construction Has Been Reduced to 1-Hour for Buildings of Type IIA, IIIA, and VA Construction Provided That the Building is Equipped with a Fire Sprinkler System and Does Not Exceed 3 Stories in Height

Control Area 1

1-Hour Fire Separation Assembly

Control Area 2

The fire-resistance rating of floor assemblies supporting control areas in some buildings is now permitted to be 1 hour.

types of construction. Requiring such buildings to provide 2-hour floor assemblies to support control areas located above the first story was not justifiable.

The new exception is still substantially more restrictive than some legacy codes, which allowed 1-hour floor assemblies for control areas in all cases. To qualify for the new 1-hour floor assembly exception in the IFC, the building must be equipped with a fire sprinkler system, the height cannot exceed three stories, and the type of construction is limited to Types IIA, IIIA, or VA. The intent is to recognize the combination of a 1-hour floor assembly plus a sprinkler system as sufficient protection in these construction types, up to three stories. The three-story limitation correlates with the increase in the required control area fire barrier rating from 1 hour to 2 hours, which is triggered by Table 2703.8.3.2 when a building exceeds three stories in height.

Note that the new exception does nothing to address the same problem in buildings with nonrated construction, because the proponent of the code change proposal that led to this revision did not recommend any changes in that regard.

Chapter 28

Aerosols

Requirements for the display and storage of aerosols have been updated.

CHANGE TYPE. Modification

CHANGE SUMMARY. Chapter 28 has been updated to correlate with Chapter 27 and the 2002 edition of NFPA 30B. **(F168-03/04, F211-04/05)**

2006 CODE:
(Portions of Chapter 28 not reprinted next remain unchanged.)

2801.1 Scope. The provisions of this chapter, the *International Building Code,* and NFPA 30B shall apply to the manufacturing, storage and display of aerosol products ~~in addition to the requirements of Chapter 27~~. Manufacturing of aerosol products using hazardous materials shall also comply with Chapter 27.

2804.3.2.1 Chain-Link Fence Enclosures. Chain-link fence enclosures required by Table 2804.3.2 shall comply with the following:

1. **and 2.** (No change to current text)
3. Class ~~III~~, IV and high-hazard commodities shall be stored outside of the aerosol storage area and a minimum of 8 feet (2438 mm) from the fence.
4. **and 5.** (No change to current text)

2806.2 Aerosol Display and Normal Merchandising Not Exceeding 8 Feet (2438 mm) High. Aerosol display and normal merchandising not exceeding 8 ft. in height shall be in accordance with Section 2806.2.1 through 2806.2.4.

2806.2.1 Maximum Quantities in Retail Display Areas. Aerosol products in retail display areas shall not exceed quantities needed for display and normal merchandising and shall not exceed the quantities in Table 2806.2.1.

TABLE 2806.2.1 Maximum Quantities of Level 2 and 3 Aerosol Products in Retail Display Areas

	Maximum Net Weight per Floor (pounds)a, b		
Floor	Unprotected e,a	Protected in accordance with Section 2806.2 a,c,d	Protected in accordance with Section 2806.3 c
Basement	Not allowed	500	500
Ground	2500	10,000	10,000
Upper	500	2000	Not allowed

For SI: 1 pound = 0.454 kg, 1 square foot = 0.0929 m^2

a. The total quantity shall not exceed 1000 pounds net weight in any one 100-square foot retail display area.
b. ~~When packaged, stored and protected in accordance with NFPA 30B, quantity limits shall be limited to those specified in NFPA 30B.~~
~~c.~~ b. Per 25,000-square-foot retail display area.
~~d.~~ c. Minimum ordinary hazard Group 2 wet-pipe automatic sprinkler system through the retail sales occupancy.

~~**2806.4**~~ **2806.2.2 Display of Containers.**   (No change to current text)

~~**2806.5**~~ **2806.2.3 Combustible Cartons.**   Aerosol products located in retail display areas shall be removed from combustible cartons.

Exceptions:
1. (No change to current text)
2. When the display area is protected in accordance with Tables ~~4–3~~ 6.3.2.7(a) through 6.3.2.7(l) of NFPA 30B, storage of aerosol products in combustible cartons is allowed.

~~**2806.6 Aisles.**   Aisles not less than 4 feet (1219 mm) in width shall be maintained on three sides of a retail display area containing aerosol products.~~

~~**2806.7**~~ **2806.2.4 Retail Display Automatic Sprinkler System.** (No change to current text.)

~~**2806.8 Storage Automatic Fire Extinguishing System.**   When the height of storage or display exceeds the limits in Section 2806.4, the design of the automatic sprinkler system shall be in accordance with NFPA 30B.~~

2806.3 Aerosol Display and Normal Merchandising Exceeding 8 Feet (2.4 m) High.   Aerosol display and merchandising exceeding 8 ft in height shall be in accordance with Section 2806.3.

2806.3.1 Maximum Quantities in Retail Display Areas.   Aerosol products in retail display areas shall not exceed quantities needed for display and normal merchandising and shall not exceed the quantities in Table 2806.2, with fire protection in accordance with Section 2806.3.2.

2806.3.2 Automatic Sprinkler Protection.   Aerosol display and merchandising areas shall be protected by an automatic sprinkler system based on the requirements set forth in NFPA 30B, Tables 6.3.2.7(a) through 6.3.2.7(l) and the following:

1. Protection shall be based on the highest level of aerosol product in the array and the packaging method of the storage located more than 6 ft. (1.8 m) above the finished floor.
2. When using the cartoned aerosol tables in NFPA 30B, uncartoned or display cut Level 2 and 3 aerosols shall be permitted not more than 6 ft. (1.8 m) above the finished floor.
3. The design area for Level 2 and Level 3 aerosols shall extend not less than 20 ft. (6 m) beyond the Level 2 and Level 3 aerosol display and merchandising areas.
4. Where ordinary and high temperature ceiling sprinkler systems are adjacent to each other, noncombustible draft curtains shall be installed at the interface.

Chapter 28 continues

Chapter 28 continued

2806.3.3 Separation of Level 2 and Level 3 Aerosol Areas.

1. Level 2 and Level 3 aerosol display and merchandising areas shall be separated from each other by not less than 25 ft. (7.6 m). Also see Table 2806.2.1.

2. Level 2 and Level 3 aerosol display and merchandising areas shall be separated from flammable and combustible liquids storage and display areas by one or combination of the following:

   2.1. Segregating areas from each other by horizontal distance of not less than 25 ft. [7.6 m].

   2.2. Isolating areas from each other by a noncombustible partition extending not less than 18 in. above the merchandise.

   2.3. In accordance with Section 2806.5.

3. When item 2.2 above is used to separate Level 2 or Level 3 aerosols from flammable or combustible liquids, and the aerosol products are located within 25 ft. (7.6 m) of flammable or combustible liquids, the area below the noncombustible partition shall be liquid-tight at the floor to prevent spilled liquids from flowing beneath the aerosol products.

2806.3 2806.4 Maximum Quantities in Storage Areas. Aerosol products in storage areas adjacent to retail display areas shall not exceed the quantities in Table 2806.3 2806.4.

TABLE 2806.3 2806.4 Maximum Storage Quantities for Storage Areas Adjacent to Retail Display of Level 2 and Level 3 AEROSOLS

| | Maximum Net Weight per Floor (Pounds) | | |
| | | Separate[d] | |
Floor	Unseparated[a,b]	Storage Cabinets[b]	1-hour Occupancy Separation
Basement	Not allowed	Not allowed	Not allowed
Ground	2500	5000	In accordance with Sections 4 3. 4.2 and 4 3.4.3 6.3.4.3 and 6.3.4.4 of NFPA 30B
Upper	500	1000	In accordance with Sections 4 3.4.2 and 4 3.4.3 6.3.4.3 and 6.3.4.4 of NFPA 30B

For SI: 1 pound = 0.454 kg, 1 square foot = 0.0929m^2

a. and b. (No change to current text)

2806.5 Special Protection Design for Level 2 and Level 3 Aerosols Adjacent to Flammable and Combustible Liquids in Double Row Racks.

The display and merchandising of Level 2 and Level 3 Aerosols adjacent to flammable and combustible liquids in double row racks shall be in accordance with Section 2806.5 or 2806.3.3.

2806.5.1 Fire Protection.

Fire protection for the display and merchandising of Level 2 and Level 3 aerosols in double-racks shall be in accordance with Table 7.4.1 and Figure 7.4.1 of NFPA 30B.

2806.5.2 Cartoned Products. Level 2 and Level 3 aerosols displayed or merchandised more than 8 ft (2.4 m) above the finished floor shall be in cartons.

2806.5.3 Shelving. Shelving in racks shall be limited to wire mesh shelving having uniform openings not more than 6 inches (152 mm) apart, with the openings comprising at least 50 percent of the overall shelf area.

2806.5.4 Aisles. Racks shall be arranged so that aisles not less than 7 ½ ft (2286 mm) wide are maintained between rows of racks and adjacent solid piled or palletized merchandise.

2806.5.5 Flue Spaces.
   1. Transverse flue spaces. Nominal 3 in. (76 mm) transverse flue spaces shall be maintained between merchandise and rack uprights.
   2. Longitudinal flue spaces. Nominal 6 in. (152 mm) longitudinal flue spaces shall be maintained.

2806.5.6 Horizontal Barriers. Horizontal barriers constructed of minimum ⅜ inch (10 mm) plywood or minimum 0.034-inch (0.086 mm) (No. 22 gauge) sheet metal shall be provided and located in accordance with Table 7.4.1 and Figure 7.4.1 of NFPA 30B when in-rack sprinklers are installed.

2806.5.7 Class I, II, III, IV and Plastic Commodities. Class I, II, III, IV and plastic commodities located adjacent to Level 2 and Level 3 aerosols shall be protected in accordance with NFPA 13.

2806.5.8 Flammable and Combustible Liquids. Class I, II, III-A and III-B Liquids shall be allowed to be located adjacent to Level 2 and Level 3 aerosol products when the following conditions are met:

   1. Class I, II, III-A and III-B Liquid Containers: Containers for Class I, II, III-A and III-B Liquids shall be limited to 1.06-gallon (4-liter) metal relieving and non-relieving style containers and 5.3-gallon (20-liter) metal relieving style containers.
   2. Fire Protection for Class I, II, III-A and III-B Liquids: Fire sprinkler protection for Class I, II, III-A and III-B Liquids shall be in accordance with Chapter 34.

CHANGE SIGNIFICANCE. Chapter 28 has been updated to correlate with Chapter 27 and the 2002 edition of NFPA 30B. The change to Section 2801.1 clarifies that storage and display of aerosols is not regulated by Chapter 27.

Specific justifications for some of the other changes are as follows:

   1. The 2003 edition text of Section 2804.3.2.1 was based on NFPA 30B, 1990 edition. Deletion of Class III commodities cor-

Chapter 28 continues

Chapter 28 continued

relates this section with NFPA 30B, 2002 edition, Section 6.3.5.3.2, Item 3, and fixes what was believed to have been an error in prior editions of the IFC.

2. Section 2806 has been reorganized and updated from correlation with the 1990 edition of NFPA 30B to the 2002 edition, Section 7.1.

3. Table 2806.2.1, Footnote "b," was deleted in favor of incorporating the referenced NFPA 30B provisions into Chapter 28. Sections 2806.2 and 2806.3 now specify the applicable requirements. Provisions in Section 2806.3 are correlated with NFPA 30B, 2002 edition, Sections 7.2.4.1, 7.2.5, 7.2.6, 7.2.6.1, and Tables 6.3.2.7(a) through 6.3.2.7(l).

4. Section 2806.8 was deleted on the basis of reorganization of the requirements and the addition of the new Sections 2806.4 and 2806.5.

5. Aisle requirements are now located in Section 2806.5.4.

6. Table 2806.4 has been updated from correlating with the 1990 edition of NFPA 30B to the 2002 edition, Sections 6.3.4.3 and 6.3.4.4.

7. Section 2806.5 has been added to correlate with Section 7.4 of NFPA 30B, 2002 edition.

CHANGE TYPE.   Modification and Addition

CHANGE SUMMARY.   The definition of "combustible fibers" has been modified to recognize the uniqueness of densely packed baled cotton, and new definitions of "baled cotton" and "baled cotton, densely packed" have been added. **(F204-04/05)**

2006 CODE:

2902.1 Combustible Fibers.   Readily ignitable and free-burning ~~fibers~~ materials in a fibrous or shredded form, such as cocoa fiber, cloth, cotton, excelsior, hay, hemp, henequen, istle, jute, kapok, oakum, rags, sisal, Spanish moss, straw, tow, wastepaper, certain synthetic fibers or other like materials. This definition does not include densely packed baled cotton.

Baled Cotton.   A natural seed fiber wrapped in and secured with industry-accepted materials, usually consisting of burlap, woven polypropylene, polyethylene or cotton, or sheet polyethylene, and secured with steel, synthetic, or wire bands, or wire; also includes linters (lint removed from the cottonseed) and motes (residual materials from the ginning process).

Baled Cotton, Densely Packed.   Cotton, made into banded bales, with a packing density of at least 22 pounds per cubic foot (360 kg/m^3), and dimensions complying with the following: a length of 55 inches (1397 mm), a width of 21 inches (533.4 mm) and a height of 27.6 to 35.4 inches. (701 to 899 mm).

Section 2902.1 continues

New definitions for baled cotton and densely packed baled cotton have been added to assist in determining whether cotton must comply with combustible fiber requirements.

Section 2902.1 continued

CHANGE SIGNIFICANCE. These revisions are part of a larger package of changes that exclude densely packed baled cotton from hazardous materials regulations and Group H occupancy requirements (see also the discussion of revisions to Table 2703.1.1[1] herein). The new and revised definitions help to delineate the different types of cotton bales, and they help to differentiate densely packed baled cotton from other combustible fibers.

Changes in the regulations for densely packed baled cotton were based on fire tests demonstrating that the burning characteristics of baled cotton are not severe enough to warrant regulating this material in the same manner as other combustible fibers. The cotton industry reports that over 99% of all U.S. cotton is currently pressed and stored in densely packed bales meeting the weight and dimension requirements of ISO 8115, *Cotton Bales–Dimensions and Density.*

One reason that the cotton industry has chosen to use densely packed bales is that they are very difficult to ignite, which allows them to be transported without being labeled as "flammable solids" or "dangerous goods." According to the proponent of this code change, the U.S. Department of Transportation, the U.S. Coast Guard, the United Nations, and the International Maritime Organization all exclude densely packed baled cotton from classification as a hazardous material.

CHANGE TYPE.   Addition

CHANGE SUMMARY.   Requirements have been added to mandate pressure-relief devices on containers, cylinders, and tanks containing compressed gases, unless such vessels are specifically exempted based on their design standards. **(F214-04/05)**

2006 CODE: **3003.3 Pressure-Relief Devices.**   Pressure-relief devices shall be in accordance with Sections 3003.3.1 through 3003.3.5.

3003.3.1 Where Required.   Pressure-relief devices shall be provided to protect containers, cylinders, and tanks containing compressed gases from rupture in the event of overpressure.

> **Exception:** Cylinders, containers, and tanks when exempt from the requirements for pressure-relief devices specified by the standards of design listed in Section 3003.3.2.

3003.3.2 Design.   Pressure-relief devices to protect containers shall be designed and provided in accordance with CGA S-1.1, CGA S-1.2, CGA S-1.3 or the ASME *Boiler and Pressure Vessel Code,* Section VIII as applicable.

3003.3.3 Sizing.   Pressure-relief devices shall be sized in accordance with the specifications to which the container was fabricated and to material-specific requirements as applicable.

3003.3 continues

3003.3

Pressure Relief for Compressed Gas Containers, Cylinders, and Tanks

This pair of pressure relief valves protects a high-pressure ammonia vessel. The device below the relief valves permits selection of either relief valve to remain in service, individually, so that the other can be easily removed for service or replacement without having to shutdown the protected vessel.

Provisions have been added to the IFC to address the need for pressure-relief devices on compressed gas containers, cylinders, and tanks. This photo illustrates tubing that is inadequately sized to satisfy pressure-relief requirements.

3003.3 continued

3003.3.4 Arrangement. Pressure-relief devices shall be arranged to discharge upward and unobstructed to the open air in such a manner as to prevent any impingement of escaping gas upon the container, adjacent structures, or personnel.

> **Exception:** DOTn specification containers having an internal volume of 30 cubic feet (0.855 m^3) or less.

3003.3.5 Freeze Protection. Pressure-relief devices or vent piping shall be designed or located so that moisture cannot collect and freeze in a manner that would interfere with the operation of the device.

CHANGE SIGNIFICANCE. Most pressure vessels are equipped with pressure-relief devices to relieve excess pressure when the vessel is exposed to fire or other abnormal thermal conditions. The design standards for pressure-relief devices (PRDs) are set forth in Chapter 45.

Although requirements for pressure relief have previously been provided in the IFC for cryogenic fluids, comparable requirements for compressed gases have been notably absent. Considering the risk to life and property created by compressed gas vessels without proper venting, it was deemed important that these provisions be added to the IFC rather than simply relying on compliance with industry standards. The new IFC provisions are consistent with good practice standards utilized by the gas supply industry.

CHANGE TYPE. Addition

CHANGE SUMMARY. A new section has been added to permit and prescribe regulations for gas vaults. The provisions are modeled after liquid storage vault provisions, which have been included in the IFC and legacy codes for many years. **(F169-03/04)**

2006 CODE: 3003.16 Vaults. Generation, compression, storage and dispensing equipment for compressed gases shall be allowed to be located in either above- or below-grade vaults complying with Sections 3003.16.1 through 3003.16.14.

3003.16.1 Listing Required. Vaults shall be listed by a nationally recognized testing laboratory.

> **Exception:** Where approved by the fire code official, below-grade vaults are allowed to be constructed on site, provided that the design is in accordance with the *International Building Code* and that special inspections are conducted to verify structural strength and compliance accordance with Section 1707 of the *International Building Code.* Installation plans for below-grade vaults that are constructed on site shall

3003.16 continues

3003.16
Vaults for Compressed Gases

Vaults such as this one, which have traditionally been used for containment of flammable and combustible liquids, are now permitted to contain equipment used to generate, compress, store, or dispense compressed gases.

3003.16 continued be prepared by, and the design shall bear the stamp of, a professional engineer. Consideration shall be given to soil and hydrostatic loading on the floors, walls, and lid; anticipated seismic forces; uplifting by ground water or flooding; and to loads imposed from above, such as traffic and equipment loading on the vault lid.

3003.16.2 Design and Construction.   The vault shall completely enclose generation, compression, storage, or dispensing equipment located in the vault. There shall be no openings in the vault enclosure except those necessary for vault ventilation and access, inspection, filling, emptying, or venting of equipment in the vault. The walls and floor of the vault shall be constructed of reinforced concrete at least 6 inches (152 mm) thick. The top of an above-grade vault shall be constructed of noncombustible material and shall be designed to be weaker than the walls of the vault to ensure that the thrust of any explosion occurring inside the vault is directed upward.

The top of an at- or below-grade vault shall be designed to relieve safely or contain the force of an explosion occurring inside the vault. The top and floor of the vault and the tank foundation shall be designed to withstand the anticipated loading, including loading from vehicular traffic, where applicable. The walls and floor of a vault installed below grade shall be designed to withstand anticipated soil and hydrostatic loading. Vaults shall be designed to be wind and earthquake resistant, in accordance with the *International Building Code.*

3003.16.3 Secondary Containment.   Vaults shall be substantially liquid tight and there shall be no backfill within the vault. The vault floor shall drain to a sump. For premanufactured vaults, liquid tightness shall be certified as part of the listing provided by a nationally recognized testing laboratory. For field-erected vaults, liquid tightness shall be certified in an approved manner.

3003.16.4 Internal Clearance.   There shall be sufficient clearance within the vault to allow for visual inspection and maintenance of equipment in the vault.

3003.16.5 Anchoring.   Vaults and equipment contained therein shall be suitably anchored to withstand uplifting by groundwater or flooding. The design shall verify that uplifting is prevented even when equipment within the vault is empty.

3003.16.6 Vehicle Impact Protection.   Vaults shall be resistant to damage from the impact of a motor vehicle, or vehicle impact protection shall be provided in accordance with Section 312.

3003.16.7 Arrangement.   Equipment in vaults shall be listed or approved for above-ground use. Where multiple vaults are provided, adjacent vaults shall be allowed to share a common wall. The common wall shall be liquid- and vapor-tight and shall be designed to withstand the load imposed when the vault on either side of the wall is filled with water.

3003.16.8 Connections. Connections shall be provided to permit the venting of each vault to dilute, disperse, and remove vapors prior to personnel entering the vault.

3003.16.9 Ventilation. Vaults shall be provided with an exhaust ventilation system installed in accordance with Section 2704.3. The ventilation system shall operate continuously or be designed to operate upon activation of the vapor or liquid detection system. The system shall provide ventilation at a rate of not less than 1 cubic foot per minute (cfm) per square foot of floor area [0.00508 m^3/(s m^2)], but not less than 150 cfm [0.071 m^3/(s m^2)]. The exhaust system shall be designed to provide air movement across all parts of the vault floor for gases having a density greater than air and across all parts of the vault ceiling for gases having a density less than air. Supply ducts shall extend to within 3 inches (76 mm), but not more than 12 inches (305 mm), of the floor. Exhaust ducts shall extend to within 3 inches (76 mm), but not more than 12 inches (305 mm) of the floor or ceiling, for heavier-than-air or lighter-than-air gases, respectively. The exhaust system shall be installed in accordance with the *International Mechanical Code.*

3003.16.10 Monitoring and Detection. Vaults shall be provided with approved vapor and liquid detection systems and equipped with on-site audible and visual warning devices with battery backup. Vapor detection systems shall sound an alarm when the system detects vapors that reach or exceed 25 percent of the lower explosive limit (LEL) or one-half the immediately dangerous to life and health (IDLH) concentration for the gas in the vault. Vapor detectors shall be located no higher than 12 inches (305 mm) above the lowest point in the vault for heavier-than-air gases and no lower than 12 inches (305 mm) below the highest point in the vault for lighter-than-air gases. Liquid detection systems shall sound an alarm upon detection of any liquid, including water. Liquid detectors shall be located in accordance with the manufacturers' instructions. Activation of either vapor or liquid detection systems shall cause a signal to be sounded at an approved, constantly attended location within the facility served by the tanks or at an approved location. Activation of vapor detection systems shall also shut off gas-handling equipment in the vault and dispensers.

3003.16.11 Liquid Removal. Means shall be provided to recover liquid from the vault. Where a pump is used to meet this requirement, it shall not be permanently installed in the vault. Electric-powered portable pumps shall be suitable for use in Class I, Division 1 locations, as defined in the ICC *Electrical Code.*

3003.16.12 Relief Vents. Vent pipes for equipment in the vault shall terminate at least 12 feet (3658 mm) above ground level.

3003.16.13 Accessway. Vaults shall be provided with an approved personnel accessway with a minimum dimension of 30 inches (762 mm) and with a permanently affixed, nonferrous ladder.

3003.16 continues

3003.16 continued

Accessways shall be designed to be nonsparking. Travel distance from any point inside a vault to an accessway shall not exceed 20 feet (6096 mm). At each entry point, a warning sign indicating the need for procedures for safe entry into confined spaces shall be posted. Entry points shall be secured against unauthorized entry and vandalism.

3003.16.14 Classified Area. The interior of a vault containing a flammable gas shall be designated a Class I, Division 1 location, as defined in the ICC *Electrical Code.*

CHANGE SIGNIFICANCE. This new section extends the permissible use of vaults to include gaseous materials. The code has previously permitted the use of vaults for protection of hazardous liquids (see Section 3404.2.8), and these provisions have now been extended to allow the use of vaults for compressed gases. Protection features applicable to vault design have been modified to accommodate gases, as opposed to liquids.

Because the code allows gases to be stored and handled inside of buildings, it is logical to permit gas storage and handling equipment to be located in vaults, given that a vault is unoccupied and poses fewer possible accident scenarios.

Chapter 33
Definition of Explosives

An example of a Type 2 magazine for storage of explosive materials.

CHANGE TYPE. Modification

CHANGE SUMMARY. The approach to separating explosive materials from exposures has been significantly revised to better accommodate industrial explosives. Terminology in Chapter 33 is now better correlated with latest regulations and good practice standards published by the United States Department of Defense and Institute of Makers of Explosives. **(F170-03/04)**

2006 CODE: **3302.1 Definitions.** The following words and terms shall, for the purposes of this chapter and as used elsewhere in this code, have the meanings shown herein.

Operating Line. A group of buildings, facilities, or workstations so arranged as to permit performance of the steps in the manufacture of an explosive or in the loading, assembly, modification, and maintenance of ammunition or devices containing explosive materials.

Quantity-Distance (Q-D). The quantity of explosive material and separation distance relationships providing protection. These relationships are based on levels of risk considered acceptable for the stipulated exposures and are tabulated in the appropriate Q-D tables. The separation distances specified afford less than absolute safety.

Minimum Separation Distance (D_0). The minimum separation distance between adjacent buildings occupied in conjunction with the manufacture, transportation, storage, or use of explosive materials where one of the buildings contains explosive materials and the other building does not.

Intraline Distance (ILD) or Intraplant Distance (IPD). The distance to be maintained between any two operating buildings on an explosives manufacturing site at least one of which contains or is designed to contain explosives or the distance between a magazine and an operating building.

Inhabited Building Distance (IBD). The minimum separation distance between an operating building or magazine containing explosive materials and an inhabited building or site boundary.

Intermagazine Distance (IMD). The minimum separation distance between magazines.

Public Traffic Route (PTR). Any public street, road, highway, navigable stream, or passenger railroad that is used for through traffic by the general public.

CHANGE SIGNIFICANCE. The approach to separating explosive materials from exposures has been significantly revised. The provisions now better accommodate industrial/commercial explosives.

Chapter 33 continues

Chapter 33 continued

To execute the changes, several new definitions have been added to Chapter 33 to correlate with common industry terminology. The terms that have been added are consistent with terminology used by the U.S. Department of Defense (DOD) and the Institute of Makers of Explosives (IME), although there are subtle differences with respect to DOD to account for application to commercial operations.

The term "Quantity-Distance" (Q-D) reflects the relationship between a quantity of explosive material and the minimum separation distances required. The use of Q-D relationships to establish building siting is a fundamental aspect of site planning and occupancy of buildings used to contain explosive materials. Distances required vary depending on the sensitivity of the receptor and are generally required to be greater where the public or those not engaged in the manufacturing process are involved. When adequate separation distances cannot be provided, quantities must be reduced.

The sub-elements of the Q-D definition are common industry acronyms that should appear on building and site plans that are used to document site layouts. Acronyms for these sub-elements, which include ILD, IPD, IBD, and IMD, are typically used in the explosives industry to describe the distance separating explosives from receptors such as inhabited buildings, public traffic or transportation routes (highways), and other storage (magazines). The acronyms are now included in the various Q-D tables in the code, and definitions have been provided in Section 3302.

In the IFC, the term "intraline distance" (ILD), often applied to military facilities, is used synonymously with "intraplant distance" (IPD), often applied to commercial facilities. By equating the terms ILD and IPD, the code focuses on the characteristics of explosive materials without regard to commercial or military application.

The term "operating line" has been defined in relationship to operating buildings, which may be grouped in a manner that creates a manufacturing process.

Notwithstanding the distances established by the Q-D tables, there are occasions where ancillary buildings not containing explosive materials are needed nearby. The new term "D_0" reflects a fire separation distance that is conceptually comparable to the approach used by the IBC to establish minimum separation distances for "detached buildings."

CHANGE TYPE. Modification

CHANGE SUMMARY. Separation distances for explosives from exposures and property lines, including the minimum separation distance (D_o), intraplant distance (IPD), inhabited building distance (IBD), and public transportation route (PTR) distance, are based on the explosive's hazard classification and the quantity of explosive material stored in buildings and magazines. **(F170-03/04)**

2006 CODE: *(Only code sections that were changed with respect to separation distances are shown next)*

3301.8.1 Quantity of Explosives. The quantity-distance tables in Sections 3304.5 and 3305.3 shall be used to provide ~~appropriate~~ the minimum separation distances from potential explosion sites as set forth in Tables 3301.8.1(1) through 3301.8.1(3). The classification of the explosives and the weight of the explosives are primary characteristics governing the use of these tables. The net explosive weight shall be determined in accordance with Sections 3301.8.1.1 through 3301.8.1.4.

Chapter 33

Separation Distances for Explosives

The intermagazine distance (IMD) between these magazines is inadequate to allow exposure separation distances to be based on the quantity of explosives in any individual magazine. Accordingly, the IFC requires that exposure separation distances be based on the sum of the net explosive weight in all magazines.

TABLE 3301.8.1(1) Application of Separation Distance (Q-D) Tables Division 1.1, 1.2 and 1.5 Explosives[a,b,c]

Item	Magazine	Q-D	Operating Building	Q-D	Inhabited Building	Q-D	Public Traffic Route	Q-D
Magazine	Table 3304.5.2(2)	IMD	Table 3305.3	ILD or IPD	Table 3304.5.2(2)	IBD	Table 3304.5.2(2)	PTR
Operating Building	Table 3304.5.2(2)	ILD or IPD	Table 3305.3	ILD or IPD	Table 3304.5.2(2)	IBD	Table 3304.5.2(2)	PTR
Inhabited Building	Table 3304.5.2(2)	IBD	Table 3304.5.2(2)	IBD	NA	NA	NA	NA
Public Traffic Route	Table 3304.5.2(2)	PTR	Table 3304.5.2(2)	PTR	NA	NA	NA	NA

a. The minimum separation distance (D_0) shall be a minimum of 60 feet. Where a building or magazine containing explosives is barricaded, the minimum distance shall be 30 feet.

b. Linear interpolation between tabular values in the referenced Q-D tables shall not be allowed. Non-linear interpolation of the values shall be allowed subject to an approved technical opinion and report prepared in accordance with Section 104.7.2.

c. For definitions of Quantity-Distance abbreviations IBD, ILD, IMD, IPD, and PTR, see Section 3302.1.

Chapter 33 continues

Chapter 33 continued

TABLE 3301.8.1(2)   Application of Separation Distance (Q-D) Tables Division 1.3 Explosives[a,b,c]

Item	Magazine	Q-D	Operating Building	Q-D	Inhabited Building	Q-D	Public Traffic Route	Q-D
Magazine	Table 3304.5.2(3)	IMD	Table 3304.5.2(3)	ILD or IPD	Table 3304.5.2(3)	IBD	Table 3304.5.2(3)	PTR
Operating Building	Table 3304.5.2(3)	ILD or IPD	Table 3304.5.2(3)	ILD or IPD	Table 3304.5.2(3)	IBD	Table 3304.5.2(3)	PTR
Inhabited Building	Table 3304.5.2(3)	IBD	Table 3304.5.2(3)	IBD	NA	NA	NA	NA
Public Traffic Route	Table 3304.5.2(3)	PTR	Table 3304.5.2(3)	PTR	NA	NA	NA	NA

a. The minimum separation distance (D_0) shall be a minimum of 50 feet.

b. Linear interpolation between tabular values in the referenced Q-D table shall be allowed.

c. For definitions of Quantity-Distance abbreviations IBD, ILD, IMD, IPD, and PTR, see Section 3302.1.

TABLE 3301.8.1(3)   Application of Separation Distance (Q-D) Tables Division 1.4 Explosives[a,b,c]

Item	Magazine	Q-D	Operating Building	Q-D	Inhabited Building	Q-D	Public Traffic Route	Q-D
Magazine	Table 3304.5.2(4)	IMD	Table 3304.5.2(4)	ILD or IPD	Table 3304.5.2(4)	IBD	Table 3304.5.2(4)	PTR
Operating Building	Table 3304.5.2(4)	ILD or IPD	Table 3304.5.2(4)	ILD or IPD	Table 3304.5.2(4)	IBD	Table	PTR
Inhabited Building	Table 3304.5.2(4)	IBD	Table 3304.5.2(4)	IBD	NA	NA	NA	NA
Public Traffic Route	Table 3304.5.2(4)	PTR	Table 3304.5.2(4)	PTR	NA	NA	NA	NA

a. The minimum separation distance (D_0) shall be a minimum of 50 feet.

b. Linear interpolation between tabular values in the referenced Q-D table shall not be allowed.

c. For definitions of Quantity-Distance abbreviations IBD, ILD, IMD, IPD, and PTR, see Section 3302.1.

3304.5.2 Outdoor Magazines. All outdoor magazines other than Type 3 shall be located so as to comply with Table 3304.5.2(2), Table 3304.5.2(3) or Table 3304.5.2.4(4) as set forth in Tables ~~3304.5.2(1)~~ 3301.8.1(1) through 3301.8.1(3). Where a magazine or group of magazines, as described in Section 3304.5.2.2, contains different classes of explosive materials, and Division 1.1 materials are present, the required separations for the magazine or magazine group as a whole shall comply with Table 3304.5.2(2).

3304.5.2.1 Separation. Where two or more storage magazines are located on the same property, each magazine shall comply with the minimum distances specified from inhabited buildings, public transportation routes, and operating buildings. Magazines shall be separated from each other by not less than the intermagazine distances (IMD) shown for separation of magazines.

3304.5.2.2 Grouped Magazines.   Where two or more magazines are separated from each other by less than the intermagazine distances (IMD), then such magazines, as a group, shall be considered as one magazine, and the total quantity of explosive materials stored in the group shall be treated as if stored in a single magazine. The location of the group of magazines shall comply with the intermagazine distances (IMD) specified from other magazines or magazine groups, inhabited buildings (IBD), public transportation routes (PTR), and operating buildings (ILD) or (IPD) as required.

TABLE 3304.5.2(1)   Application of Separation Distance Table*

TABLE 3304.5.2(2)   Table of Distances (Q-D) for Buildings Containing Explosives—Division 1.3 Mass-Fire Hazard[a, b, c]

Quantity of Division 1.3 Explosives (Net Explosives Weight)		Distances in Feet			
Pounds over	Pounds not over	Inhabited Buildings Distance (IBD)	Passenger railways and public highways Distance to Public Traffic Route (PTR)	Magazines and operating buildings Intermagazine Distance (IMD)	Intraline Distance (ILD) or Intraplant Distance (IPD)
0	1000	75	75	50	50
1000	5000	115	115	75	75
5000	10,000	150	150	100	100
10,000	20,000	190	190	125	125
20,000	30,000	215	215	145	145
30,000	40,000	235	235	155	155
40,000	50,000	250	250	165	165
50,000	60,000	260	260	175	175
60,000	70,000	270	270	185	185
70,000	80,000	280	280	190	190
80,000	90,000	295	295	195	195
90,000	100,000	300	300	200	200
100,000	200,000	375	375	250	250
200,000	300,000	450	450	300	300

(Current notes and footnotes do not change.)

TABLE 3304.5.2(3)   Table of Distances (Q-D) for Buildings Containing Explosives—Division 1.4[c]

Quantity of Division 1.4 Explosives (pounds) (Net Explosives Weight)		Distances in Feet			
Pounds over	Pounds not over	Inhabited Buildings Distance IBD	Passenger railways and public highways Distance to Public Traffic Route (PTR)	From Above-ground Magazines and operating buildings[a, b] Intermagazine Distance[a, b] (IMD)	Intraline Distance (ILD) or Intraplant Distance[a] (IPD)
50	Not limited	100	100	50	50

(Current notes and footnotes do not change.)

Chapter 33 continues

Chapter 33 continued

TABLE 3305.3   Minimum <u>Intraline</u> (Intraplant) Separation Distances (<u>ILD or IPD</u>) Between Barricaded Operating Buildings Containing Explosives—Division 1.1, 1.2 or 1.5—Mass Explosion Hazard[a]

<u>Net</u> Explosives <u>Weight</u>		<u>Net</u> Explosives <u>Weight</u>			
		~~Minimum distance~~ Intraline Distance (<u>ILD)</u> or Intraplant			~~Minimum distance~~ Intraline Distance (<u>ILD)</u> or Intraplant
Pounds over	Pounds not over	Distance (IPD) (feet)	Pounds over	Pounds not over	Distance (IPD) (feet)

(Portions of table and footnotes not shown do not change.)

CHANGE SIGNIFICANCE.   Minimum separation distances in Chapter 33, denoted as D_0, are established by Tables 3301.8.1(1) through 3301.8.1(3), with 50 feet being the minimum separation distance for buildings containing materials presenting mass fire and fire hazards and 60 feet being the minimum separation distance for buildings containing materials presenting mass explosion hazards. From a practical standpoint, the D_0 distances will apply only to facilities where the explosive quantities are at or near the minimums, because the required separation distances increase rapidly as explosives quantities are increased.

Section 3301.8 requires the quantity of explosives and separation distances to be in accordance with the limits set forth in the Q-D tables in Chapter 33 [Tables 3304.5.2(1), 3304.5.2(2), 3304.5.2(3), and 3305.3], as appropriate. Guidance in the use and applicability of these tables has been enhanced by deleting Table 3304.5.2(1), which was included in the 2003 edition and was limited in applicability to magazines, and by adding three new tables [3301.8.1(1), 3301.8.1(2), and 3301.8.1(3)], which deal with magazines as well as operating buildings. The three new tables are similar in application to the old Table 3304.5.2(1) in that they assist in identifying which Q-D table should be used for various situations. Again, the acronyms IPD, ILD, IMD, etc., in the tables are defined in Section 3302 under "Quantity-Distance."

The revised Chapter 33 also provides new guidance regarding permissible interpolation of distances when using the tables. In general, linear interpolation is not appropriate for materials with a mass explosion hazard (high explosives), as the effect of an explosion varies with the cube root of the distance (Table 3301.8.1[1]). Nonlinear interpolation may be performed with the proviso that the interpolation be documented in an approved "Technical Opinion and Report," as required by Section 104.7.2.

For materials with mass fire hazards shown in Table 3301.8.1(2), linear interpolation is allowed, as noted in Footnote "c" to Table 3304.5.2(2).

Interpolation of Table 3304.5.2(3) is not applicable, as the table does not provide a series of quantity ranges where interpolation could be used. It should be noted that the use of this table is restricted to "articles" (Division 1.4), including articles packaged for shipment, that are not regulated as an explosive under Bureau of Alcohol, Tobacco and Firearms regulations, or unpacked articles used in process operations that do not propagate a detonation or deflagration between articles.

It is common to group a number of magazines, often containing different materials, on manufacturing sites for the purposes of operational control and dedication to specific manufacturing lines, and two new sections have been added to Section 3304.5.2 (3304.5.2.1 and 3304.5.2.2) to provide appropriate regulations. The new provisions address the use of multiple magazines and magazines that are arranged in groups where the separation between magazines is less than the tabular distance established by the "intermagazine" distances (meaning between or among magazines). The basis for these provisions is Footnote "d" to the American Table of Distances (Table 3304.5.2[1]), except that the permissive text has been removed and requirements in the note have been divided into two distinct provisions. The new provisions are applicable to Division 1.3 and 1.4 explosive materials as well as to the materials addressed by Table 3304.5.2(1).

Tables 3304.5.2(2) and 3304.5.2(3) have been reformatted to correlate with the new acronyms established for Q-D purposes. A column has been added to Tables 3304.5.2(2) and 3304.5.2(3) to separate IMD and ILD/IPD, which had previously been combined as a single column. This change was made to clarify that the separation of magazines, as shown in the Intermagazine Distance (IMD) column, is the separation between magazines.

Chapter 33

Separation Distances for Explosives—Day Boxes and Operating Buildings

A Closed Type 3 Magazine, which is commonly referred to as a Day Box. (Photograph courtesy of the ARMAG Corporation, Bardstown, KY.)

CHANGE TYPE. Modification

CHANGE SUMMARY. Code requirements for determining the required separation distances between exposures and operating buildings where explosives are manufactured or used have been revised to provide suitable regulations for facilities handling commercial/industrial explosives. In addition, requirements for day boxes associated with operating buildings have been clarified by the addition of a new exception referencing the appropriate code provisions. **(F171-03/04)**

2006 CODE: **3304.5.1.3 Quantity Limit.** Not more than 50 pounds (23 kg) of explosives or explosive materials shall be stored within an indoor magazine.

> **Exception:** Day boxes used for the storage of in-process material in accordance with Section 3305.6.4.1.

3305.4 Separation of Manufacturing Operating Buildings from Inhabited Buildings, ~~Rights-of-Way~~ Public Traffic Routes, and Magazines. When ~~a manufacturing~~ an operating building on an explosive materials plant site is designed to contain explosive materials, such building shall be located away from inhabited buildings, public ~~highways, and passenger railways~~ traffic routes, and magazines in accordance with Table 3304.5.2(2), 3304.5.2(3), or 3304.5.2(4) as appropriate, based on the maximum quantity of explosive materials permitted to be in the building at one time. (see Section 3301.8).

> **Exception:** (No change to current text)

3305.4.1 Determination of Net Explosive Weight for Operating Buildings. In addition to the requirements of Section 3301.8 to determine the net explosive weight for materials stored or used in operating buildings, quantities of explosives materials stored in magazines located at distances less than intraline distances from the operating building shall be added to the contents of the operating building to determine the net explosive weight for the operating building.

3305.4.1.1 Indoor Magazines. The storage of explosive materials located in indoor magazines in operating buildings shall be limited to a net explosive weight not to exceed 50 pounds.

3305.4.1.2 Outdoor Magazines with a Net Explosive Weight Less than 50 Pounds. The storage of explosive materials in outdoor magazines located at less than intraline distances from operating buildings shall be limited to a net explosive weight not to exceed 50 pounds.

3305.4.1.3 Outdoor Magazines with a Net Explosive Weight Greater than 50 Pounds. The storage of explosive materials in outdoor magazines in quantities exceeding 50 pounds net explosive weight shall be limited to storage in outdoor magazines located not less than intraline distances from the operating building in accordance with Section 3304.5.2.

3305.4.1.4 Net Explosive Weight of Materials Stored in Combination Indoor and Outdoor Magazines. The aggregate quantity of explosive materials stored in any combination of indoor magazines or outdoor magazines located at less than the intraline distances from an operating building shall not exceed 50 pounds.

CHANGE SIGNIFICANCE.   Magazines may be used either as "storage" magazines or as holding areas for explosive materials awaiting same-day use in an operating building, which are commonly called "day boxes." Physically, day boxes meet the construction requirements applicable to Type 3 magazines, but day boxes are required to meet more stringent operational controls than storage magazines. For example, the quantity of explosive material in a day box is limited to that needed in a single work day.

In the 2006 edition, the quantity limits applicable to indoor storage magazines versus day boxes have been clarified by a new exception to Section 3304.5.1.3 that references Section 3305.6.4.1. Section 3304.5.1.3 ordinarily limits the quantity of explosives and explosive materials in indoor magazines to not more than 50 pounds, but this limitation governs only materials that are being stored. Materials in day box magazines that are awaiting same-day processing are governed by special quantity allowances, set forth in Section 3305.6.4.1. The new cross-reference in the exception to Section 3304.5.1.3 ensures that the special allowances for day boxes are not overlooked.

It is important to point out that magazines used for storage are regulated by the U.S. Bureau of Alcohol, Tobacco and Firearms (BATF), and the BATF requires materials in storage magazines to be identified on written logs with closely maintained inventories. However, the BATF does not regulate "in-process" materials, including materials in day boxes, so local enforcement of the IFC provisions for day boxes is important.

Provisions added to Section 3305.4.1 clarify the basis for determining the quantity of explosive material associated with manufacturing buildings. These buildings have a need for "in process" materials that are accessible and proximate to the location where they will be handled.

3308.2

Permit Application for Fireworks

CHANGE TYPE.   Modification

CHANGE SUMMARY.   A prerequisite condition has been added to the permitting requirements for fireworks displays mandating that a plan for dealing with misfires or malfunctions be documented and submitted with the permit application. **(F174-03/04)**

2006 CODE: 3308.2 Permit Application.   Prior to issuing permits for a fireworks display, plans for the display, inspections of the display site and demonstrations of the display operations shall be approved. <u>A plan establishing procedures to follow and actions to be taken in the event that a shell fails to ignite in, or discharge from, a mortar or fails to function over the fallout area or other malfunctions shall be provided to the fire code official.</u>

CHANGE SIGNIFICANCE.   The pyrotechnics industry has established procedures for responding to various types and degrees of incidents that may occur as a result of a misfire. This revision requires that such procedures be documented in a plan as part of the permit application for every display. The fire code official can then evaluate the appropriateness of plans prior to permit issuance. The terminology used is consistent with NFPA 1123, Code for Fireworks Displays.

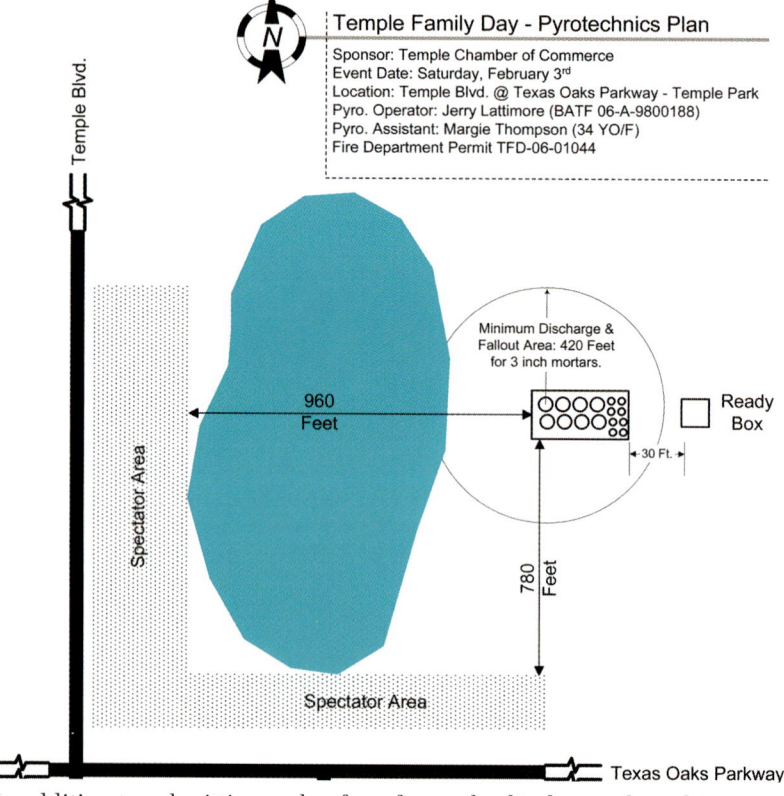

Temple Family Day - Pyrotechnics Plan
Sponsor: Temple Chamber of Commerce
Event Date: Saturday, February 3rd
Location: Temple Blvd. @ Texas Oaks Parkway - Temple Park
Pyro. Operator: Jerry Lattimore (BATF 06-A-9800188)
Pyro. Assistant: Margie Thompson (34 YO/F)
Fire Department Permit TFD-06-01044

In addition to submitting a plan for a fireworks display, such as this one, permit applicants are now required to submit a written plan documenting how malfunctioning or misfired fireworks will be safely remedied.

3308.9

Post–Fireworks Display Inspection

CHANGE TYPE. Modification

CHANGE SUMMARY. The provisions for post–fireworks display inspections have been modified to require that a report be prepared and submitted to the fire code official to detail malfunctions that occurred during the display. **(F175-03/04)**

2006 CODE: 3308.9 Post-Display Inspection. After the display, the firing crew shall conduct an inspection of the fallout area for the purpose of locating unexploded aerial shells or live components. This inspection shall be conducted before public access to the site shall be allowed. Where fireworks are displayed at night and it is not possible to inspect the site thoroughly, the operator or designated assistant shall inspect the entire site at first light. <u>A report identifying any shells that fail to ignite in, or discharge from, a mortar or fail to function over the fallout area or otherwise malfunction shall be filed with the fire code official.</u>

CHANGE SIGNIFICANCE. Post–fireworks display inspection requirements now include preparing a report of defective shells and submitting the report to the fire code official. The intent of this requirement is to help identify potential deficiencies in products or set-up procedures. The terminology used is consistent with NFPA 1123, Code for Fireworks Displays.

A report to the fire official identifying malfunctioning fireworks is now required. (Photo courtesy of Casey Lake, Pyroboy Video).

3402.1

Definition of Liquid Storage Warehouse

CHANGE TYPE.   Addition

CHANGE SUMMARY.   A definition of "liquid storage warehouse" has been added to correlate with the use of the term in several locations in the code. **(F232-04/05)**

2006 CODE: LIQUID STORAGE WAREHOUSE.   A building classified as a Group H-2 or H-3 occupancy used for the storage of flammable or combustible liquids in a closed condition.

CHANGE SIGNIFICANCE.   The term "liquid storage warehouse" was not previously defined in the IFC, even though it has been used in several locations in the code, having been inherited from the Uniform Fire Code. The new definition clarifies that the applicable occupancy classification for a liquid storage warehouse may be either Group H-2 or H-3. Which of these classifications applies depends on whether the pressure of stored liquids exceeds 15 pounds per square inch.

Normally, a liquid storage warehouse will be classified as Group H-3 because liquids will be stored in containers at atmospheric pressure; however, in some cases, liquids will be stored in pressurized vessels exceeding 15 psi, which requires an H-2 occupancy when such liquids are stored in excess of threshold quantities.

Note that flammable and combustible liquids stored as aerosols do not follow these rules for occupancy classification. Aerosols are subject to special regulations in the IBC and IFC. See the definitions of the Group H-2 and H-3 occupancy classifications in Chapter 2 of the IFC for additional information.

Rack storage of flammable liquids inside a liquid storage warehouse.

Classifications for flammable and combustible liquid storage areas progress from (1) control areas, to (2) liquid storage rooms, to (3) liquid storage warehouses, depending on the quantity of liquids being stored. Once a liquid storage warehouse is provided, the quantity of flammable and combustible liquids permitted in the building or area is not limited (see the 2006 edition change to Table 3404.3.6.3[3] discussed herein).

The term "liquid storage warehouse" is not used in the IBC, so construction requirements for liquid storage warehouses are based only on the occupancy classification, which will be Group H-2 or H-3. The IBC has no other special construction requirements. IFC requirements for liquid storage warehouses are set forth in Section 3404.3.8.

The construction requirements in Section 3404.3.8 for liquid storage warehouses are very similar to the construction requirements for liquid storage rooms, which are subject to storage quantity limits. The primary differences between liquid storage rooms and liquid storage warehouses, from a construction perspective, are (1) liquid storage warehouses require installed hose lines for firefighting, and (2) liquid storage warehouses receive an allowance to provide less exhaust ventilation than is required for liquid storage rooms.

The difference in required ventilation rates between liquid storage rooms and liquid storage warehouses resulted from a code change that took place in the 2003 edition that does not necessarily make sense, given that there is no general technical justification for permitting less ventilation where larger quantities of stored liquids will be present. The reason for the reduced ventilation rate applying only to liquid storage warehouses is that the proponent of the change to the 2003 edition did not request the ventilation rate reduction for liquid storage rooms, and no one suggested expanding the proposal in that regard during the code development process. Consequently, the approved code change impacted only liquid storage warehouses.

It should be noted that the Uniform fire and building codes and other codes include additional restrictions on the construction of liquid storage warehouses, such as increased requirements for fire-resistive separations from other occupancies, but the International Codes do not include such additional requirements.

3404.2.7.5.8 and 3406.4.6

Overfill Prevention

An overfill protection alarm for a tank.

CHANGE TYPE. Modification

CHANGE SUMMARY. Requirements to provide overfill protection for aboveground tanks have been revised, and code references to API 2350 as an applicable design and installation standard have been expanded and updated. **(F227-04/05)**

2006 CODE: 3404.2.7.5.8 Overfill Prevention. An approved means or method in accordance with Section 3404.2.9.6.6 shall be provided to prevent the overfill of all Class I, II and IIIA liquid storage tanks. Storage tanks in refineries, bulk plants, or terminals regulated by Sections 3406.4 or 3406.7 shall have overfill protection in accordance with API 2350.

> **Exception:** Outside above-ground tanks with a capacity of 1320 gallons (5000 liters) or less.

3406.4.6 Overfill Protection of Class I and II Liquids. Manual and automatic systems shall be provided to prevent overfill during the transfer of Class I and II liquids from mainline pipelines and marine vessels in accordance with API 2350.

CHANGE SIGNIFICANCE. In the 2003 edition, Section 3404.2.7.5.8 required an approved means or method of overfill prevention for all aboveground tanks containing Class I, II, or IIIA liquids, except for tanks at bulk plants and terminals, which are governed by Section 3406.4.6. In the 2006 edition, the requirements in Section 3404.2.7.5.8 have been reduced by exempting outside tanks with a capacity not exceeding 1320 gallons. Such small tanks often have fill connections that are less than 2 inches in diameter, and because overfill prevention

Above-ground storage tanks storing more than 1320 gallons of Class I and II flammable and combustible liquids require overfill protection in accordance with API 2350, *Overfill Protection for Storage Tanks in Petroleum Facilities.*

valves are not commonly available for flammable and combustible liquids in such small sizes, the cost of providing overfill prevention for these tanks was unreasonably expensive. Exclusion of tanks having a capacity not exceeding 1320 gallons is also warranted on the basis of consistency with the 2005 edition of API 2350, which exempts such tanks, and the Federal regulations for Spill Prevention Control and Countermeasures (SPCC) plans, which require plans for most sites only when the aggregate tank volume is over 1320 gallons.

Section 3404.2.7.5.8 has also been modified by adding a reference to the industry standard for overfill protection at refineries, bulk plants, and terminals (API 2350), and the reference to API 2350 in Chapter 45 has been updated to the 2005 edition.

Tanks at bulk plants and terminals are not required to follow the general rules for overfill prevention in Section 3404.2.7.5.8 because Section 3406.4.6 sets forth specific regulations that take precedence (Section 102.9 provides for specific regulations overruling general regulations). In the 2003 edition, only tanks containing Class I liquids were required to be equipped with overfill prevention controls at bulk plants and terminals. These requirements were expanded in 2006 to include tanks containing Class II liquids for consistency with API 2350–2005. However, tanks containing Class IIIA liquids at bulk plants, terminals, and refineries are still not required to be equipped with overfill prevention controls because, for these applications, Section 3404.2.7.5.8 defers to API 2350, and API 2350 exempts tanks containing Class III liquids.

Overfill monitoring equipment and other devices on top of a field-erected tank.

3404.3.1

Design, Construction, and Capacity of Containers and Portable Tanks

A "bag-in-box" container.

CHANGE TYPE.   Modification

CHANGE SUMMARY.   Class IIIB liquids are no longer subject to container construction requirements applicable to other flammable and combustible liquids. However, storage of such liquids in sprinklered buildings requires a sprinkler system design that is adequate for protection of the stored commodity, considering the packaging, the contained liquid and the arrangement of the storage array. **(F228-04/05)**

2006 CODE: 3404.3.1 Design, Construction, and Capacity of Containers and Portable Tanks.   The design, construction, and capacity of containers for the storage of ~~flammable and combustible~~ Class I, II and IIIA liquids shall be in accordance with this section and Section ~~4.2~~6.2 of NFPA 30.

CHANGE SIGNIFICANCE.

Reason: The revision to Section 3404.3.1 appears somewhat inconsequential on the surface, but it is actually quite significant. In the 2003 and prior editions of the IFC, Class IIIB liquids were required to meet the requirements for flammable and combustible liquid packaging, but in the 2006 edition, this requirement has been dropped.

Two pallet loads of 6-gallon bag-in-box containers of motor oil prior to a fire test.

The change correlates the IFC with a Tentative Interim Amendment (TIA) to NFPA 30–2003 that was approved by the NFPA Standards Council in July 2004. Because Chapter 45 of the 2006 IFC adopts the 2003 edition of NFPA 30 and because TIAs are not transcribed into the body of NFPA documents, it was important to make a correlating change to the IFC.

3404.3.1 continues

A 275-gallon bag-in-tote vessel for motor oil.

Permissible storage of bag-in-box and bag-in-tote vessels in sprinklered buildings requires a suitable basis for design of the sprinkler system. Some bag-in-box and bag-in-tote vessels meeting specific construction requirements have been qualified by full-scale fire testing as suitable for storage of motor oil in sprinklered buildings. One such test is shown here, prior to activation of the sprinkler system.

3404.3.1 continued The criteria set forth in NFPA 30 Section 6.2, including Table 6.2.3 for construction of storage vessels, have traditionally been based primarily on types of containers, intermediate bulk containers, and portable tanks that are acceptable for shipment under the rules of the U. S. Department of Transportation (DOT). However, NFPA 30 overlooked the fact that DOT does not regulate Class IIIB liquids (liquids with flash points above 200° F).

This inconsistency went unnoticed until a new packaging system for motor oil called bag-in-box, which didn't meet container construction requirements in the 2003 edition of NFPA 30, was introduced. Bag-in-box containers are comprised of a plastic bladder inside of a corrugated box. When the inconsistency between NFPA 30 and DOT regulations was brought to the attention of the NFPA 30 technical committee, the committee determined that there was no reason, from a fire-safety perspective, for prescribing special container construction requirements for Class IIIB liquids, and the requirements were dropped.

CHANGE TYPE.   Deletion

CHANGE SUMMARY.   The number of flammable and combustible liquid storage cabinets in a fire area is no longer limited, provided that aggregate quantity limits are not exceeded. **(F229-04/05)**

2006 CODE: **3404.3.2.3 Number of Storage Cabinets.** ~~Not more than three storage cabinets shall be located in a single fire area, except that in a Group F occupancy, additional cabinets are allowed to be located in the same fire area if the additional cabinets (or groups of up to three cabinets) are separated from other cabinets or groups of cabinets by at least 100 feet (30,480 mm).~~

CHANGE SIGNIFICANCE.   The IFC and legacy fire codes have permitted an increase in the maximum allowable quantity (MAQ) of flammable and combustible liquids in a control area, based on the use of liquid-storage cabinets, for many years. However, by limiting the maximum number of liquid storage cabinets permitted in a single fire area, the code unintentionally limited the ability to use the permissible MAQ quantity increase in some cases. By deleting the limit on the maximum number of storage cabinets, the code now relies directly on the MAQ limits.

It should be noted that not all occupancies are permitted unrestricted use of MAQs listed in the MAQ tables in Chapter 27. Additional regulations, restricting the permissible use of the quantities in the MAQ tables in certain use groups, are included in Sections 3404.3.4.2 and 3405.3.5.2, and these should not be overlooked.

3404.3.2.3
Number of Storage Cabinets

The previous limit of three liquid storage cabinets per fire area in most occupancies has been deleted. Now, the permissible number of cabinets is limited only by applicable material storage quantity limits.

3404.3.5.1

Basement Storage

CHANGE TYPE. Modification

CHANGE SUMMARY. Limited quantities of Class I liquids are now permitted to be stored in basements when a fire protection system is provided. Quantity limits for use of Class I liquids used in basements have been reduced. **(F230-04/05)**

2006 CODE: **3404.3.5.1 Basement Storage.** Class I liquids shall ~~not be permitted~~ be allowed to be stored in basement~~s~~ ~~areas~~ in amounts not exceeding the maximum allowable quantity per control area for use-open systems in Table 2703.1.1(1), provided that automatic suppression and other fire protection are provided in accordance with Chapter 9. Class II and IIIA liquids shall also be allowed to be stored in basements, provided that automatic suppression and other fire protection are provided in accordance with Chapter 9.

CHANGE SIGNIFICANCE. In prior editions of the IFC, Section 3404.3.5.1 prohibited storage of Class I liquids in basements; however, use of Class I liquids in basements in amounts not exceeding maximum allowable quantities for control areas (MAQs) was not prohibited and was therefore allowed (see IFC Section 3405.3.7.2). Accordingly, under the 2003 edition, if a Class I liquid was needed to support some process or use in a basement, the liquid had to be stored at the first story, or above, and then transported to the basement when needed. Because increased handling typically increases the likelihood of a

Both storage and use of Class I flammable liquids are now permitted in basement control areas equipped with an automatic fire-extinguishing system. Storage must comply with applicable housekeeping requirements.

mishap, it made little sense for the code to require liquids to be transported between floors versus simply permitting the same quantities to be stored in basements.

It is worth noting that one of the main reasons for the code previously prohibiting storage of Class I liquids in basements was concern for firefighter safety. When a fire occurs in a basement, particularly a fast-growing fire with a high rate of heat release, products of combustion travel upward and oppose firefighter access, increasing the risk of injury. This important concern has not been lost with the change in the 2006 edition, because the permissible quantity limits have not increased. On the contrary, they have actually been reduced, and fire-protection requirements have been increased because fire protection in accordance with Section 903.2.10 must now be provided for storage or use of Class I liquids in basements.

By deferring the storage quantity limit back to Table 2703.1.1(1), Footnote "b" to that table now applies, and this footnote limits the combined quantity of liquids in storage and in use to the storage MAQ. Therefore, under the 2006 code, the quantity of Class I liquid in storage and use in a basement cannot exceed the use-open MAQ, as specified in Section 3404.3.5.1. This limitation actually results in a reduction in permissible quantities for use-closed systems versus the 2003 edition, and it invokes very stringent limits on Class I liquids in storage or use-open situations.

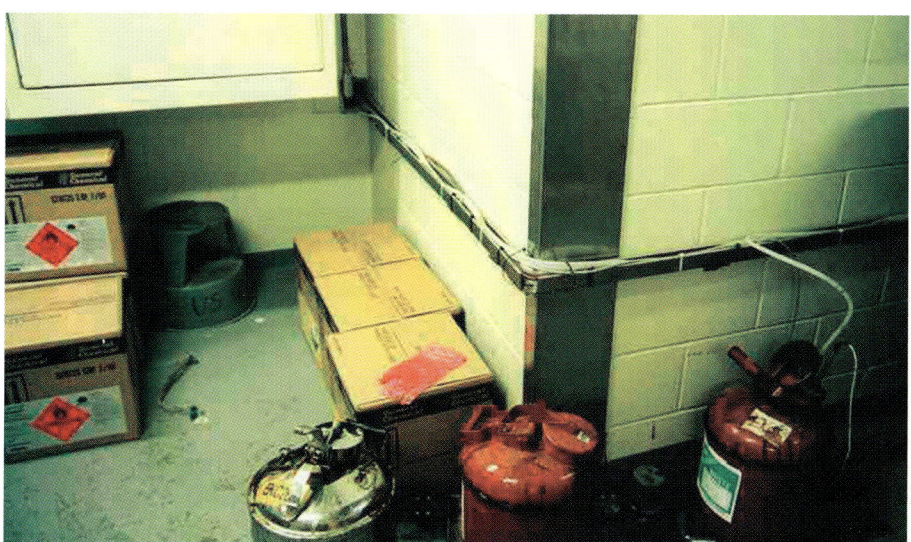

Quantities of flammable liquid exceeding 10 gallons used in basements, or elsewhere, for operations or maintenance, such as liquid in these safety cans, must be kept in a liquid storage cabinet as required by Section 3404.3.4.4 when not in use.

Table 3404.3.6.3(2)

Storage of Unsaturated Polyester Resin

Photo of one of the fire tests that justified special treatment of UPR storage.

CHANGE TYPE. Modification

CHANGE SUMMARY. Drum stacking heights permitted for unsaturated polyester resins have been increased as a result off full-scale fire testing that demonstrated the ability of properly designed fire-protection systems to satisfactorily protect such storage arrays. **(F179-03/04)**

2006 CODE:

TABLE 3404.3.6.3(2) Storage Arrangements for Palletized or Solid-Pile Storage in Liquid Storage Rooms and Warehouses

Class	Storage Level
IC	Ground floor[d]
	Upper floors
	Basements
II	Ground floor[d]
	Upper floors
	Basements

(Portions of table not shown did not change.)

d. For palletized storage of unsaturated polyester resins (UPR) in relieving-style metal containers with 50 percent or less by weight Class IC or II liquid and no Class IA or IB liquid, height and pile quantity limits shall be permitted to be 10 feet and 15,000 gallons, respectively, provided that such storage is protected by sprinklers in accordance with NFPA 30 and that the UPR storage area is not located in the same containment area or drainage path for other Class I or Class II liquids.

On the basis of full-scale fire tests, three-high drum storage of unsaturated polyester resin is now allowed when the specific IFC and NFPA 30 requirements for this material are met.

CHANGE SIGNIFICANCE. The new Footnote "d" permits special height increases for palletized arrays of certain resins in metal drums, based on the results of full-scale fire testing. Note that Table 3404.3.6.3(2) and other tables in Chapter 34 are intended to be limited to <u>metallic containers only.</u> This limitation is more clearly conveyed in NFPA 30, where more comprehensive and up-to-date tables for protection of flammable and combustible liquids can be found.

The new Footnote "d" was justified by full-scale fire tests that were sponsored by the American Composites Manufacturing Association (ACMA). These tests demonstrated that:

1. Low-flashpoint unsaturated polyester resins (UPRs) have a fire behavior more closely resembling that of Class IIIB liquids than Class I liquids, even though they are technically classified as Class I liquids in many cases based on flashpoint, and

2. Sprinkler protection schemes more closely associated with protection of Class III liquids can satisfactorily protect UPRs in relieving-style metal drums.

Phase I of the fire test program, which included three tests, was conducted at Omega Point Laboratories in San Antonio, Texas. These tests involved totally engulfing single 55-gallon relieving-style drums containing a representative UPR in a UPR liquid pool fire to determine the behavior of the commodity in extreme fire conditions. The results of Phase I tests were considered successful in demonstrating that UPR is a unique commodity that does not behave like a "typical" flammable or combustible liquid in a fire condition because of high viscosity and density. During the Phase I tests, no significant increases in internal drum pressure were observed in any case, even though drum contents completely polymerized during the tests, changing from a liquid to a dense jelly-like consistency.

Phase II tests were conducted at Southwest Research Institute (SwRI) in San Antonio, Texas, in a simulated warehouse environment and included arrangements ranging in size from a single pallet load to a 3-high pallet array. These tests successfully demonstrated that UPRs respond better to the application of sprinkler water than "typical" low-flashpoint flammable liquids. The unique fire behavior, which resembled that of a Class IIIB liquid as opposed to Class IC or Class II liquids (which were the true classifications of tested liquids based on flashpoint), can be attributed to at least two factors. First, UPRs are more viscous than typical fluids, which tends to limit the size of associated pool fires, and second, UPRs are heavier than water, which allows sprinkler discharge to cool the surface of a burning pool fire more efficiently than with typical liquids.

The above-mentioned fire testing conducted at SwRI served as the basis for changing the fire sprinkler design requirements for UPRs in the 2003 edition of NFPA 30. The design requirements in NFPA 30 permit 10-foot-high storage arrays for UPR based on protection provided by a water-only sprinkler system discharging 0.45 gpm/sq ft. This is an arrangement that achieved successful results in the SwRI tests. It is no-

Table 3404.3.6.3(2) continues

Table 3404.3.6.3(2) continued

table that the SwRI tests demonstrated that pile length and width and the associated liquid volume are not factors in determining the effectiveness of sprinkler protection for UPRs. Accordingly, the IFC permits the length and width of UPR storage arrays to parallel the limits for Class III liquids, rather than Class I or Class II liquids.

The limitations prohibiting commingling with or exposure to the drainage path of other Class I and Class II liquids are provided because the fire exposures used in the SwRI tests involved burning UPRs. These tests did not provide enough information to accurately predict what would happen if a fire involving ordinary Class I or Class II liquids were to expose piled UPR storage. Accordingly, NFPA 30 and the IFC do not permit special allowances for UPRs if there is a potential for ordinary Class I or Class II liquids to flow beneath or pool under specially protected UPR arrays.

CHANGE TYPE.   Modification

CHANGE SUMMARY.   The application of Table 3404.3.6.3(3) has been clarified by the addition of a new Footnote "a" to indicate that the quantities of flammable and combustible liquids in a liquid-storage warehouse are not limited. **(F232-04/05)**

2006 CODE:

TABLE 3404.3.6.3(3)   Storage Arrangements for Rack Storage in Liquid Storage Rooms and Warehouses

(Portions of table not shown did not change.)

a. See Section 3404.3.8.1 for unlimited quantities in liquid storage warehouses.

CHANGE SIGNIFICANCE.   The new Footnote "a" was added to help clarify that the maximum quantity-per-room limitations in the table do not apply to liquid storage warehouses. In liquid storage warehouses, only the storage height limits in the table apply. Detailed requirements for liquid storage warehouses are provided in IFC Section 3404.3.8.1.

A similar footnote is included in the 2006 and prior-edition versions of Table 3404.3.6.3(2); however, that table applies only to palletized and solid piled storage. By adding the footnote to Table 3404.3.6.3(3), racked storage in liquid storage warehouses is also addressed and correlated.

The new Footnote "a" also correlates with a new definition of "liquid storage warehouse" in Section 3402.1. Additional information on the new definition is provided in the discussion of Section 3402.1 herein.

Table 3404.3.6.3(3)

Quantity Limits for Liquid Storage Warehouses

The quantity of Class I, II, and III liquids in a liquid storage warehouse is unlimited.

3405.3.8.4

Weather Protection

CHANGE TYPE. Addition

CHANGE SUMMARY. A new section has been added to indicate that use of flammable and combustible liquids in outdoor locations under a code-compliant weather protection structure can still be considered as outdoor use. **(F180-03/04)**

2006 CODE: **3405.3.8.4 Weather Protection.** Weather protection for outdoor use shall be in accordance with Section 2705.3.9.

CHANGE SIGNIFICANCE. Storage of flammable and combustible liquids in an open area outdoors is clearly outdoor storage, and storage of flammable and combustible liquids inside of a building is clearly indoor storage. The same is clear for liquids being used. However, when an outdoor area is covered by a weather protection structure, the delineation of outdoor versus indoor becomes less clear.

In the 2006 and prior editions of the code, guidance on the classification of areas beneath weather protection structures as outdoor versus indoor areas has been provided for _storage_ of flammable and combustible liquids (by reference in Section 3404.4.7 to Section 2704.13). However, Chapter 34 did not provide a direct reference to the general hazardous materials provisions in Section 2705.3.9, which permit hazardous materials _in use_ beneath a weather protection structure to be classified as use outdoors. The new Section 3405.3.8.4 fills this gap.

IBC Section 414.6.1 prescribes the limits for weather protection structures to qualify as outdoor storage. Such structures:

1. Are permitted to have the greater of one side or 25% of the perimeter obstructed,
2. Must be of noncombustible construction, and
3. Must not exceed 1500 square feet plus applicable area increases permitted by IBC Section 506.

The requirements for outdoor use of flammable and combustible liquids under a weather protection structure have been clarified and improved.

Areas beneath weather protection structures meeting these requirements are classified as outdoor areas. Areas beneath weather protection structures not meeting these limitations must be classified and regulated as indoor areas and assigned an appropriate occupancy classification.

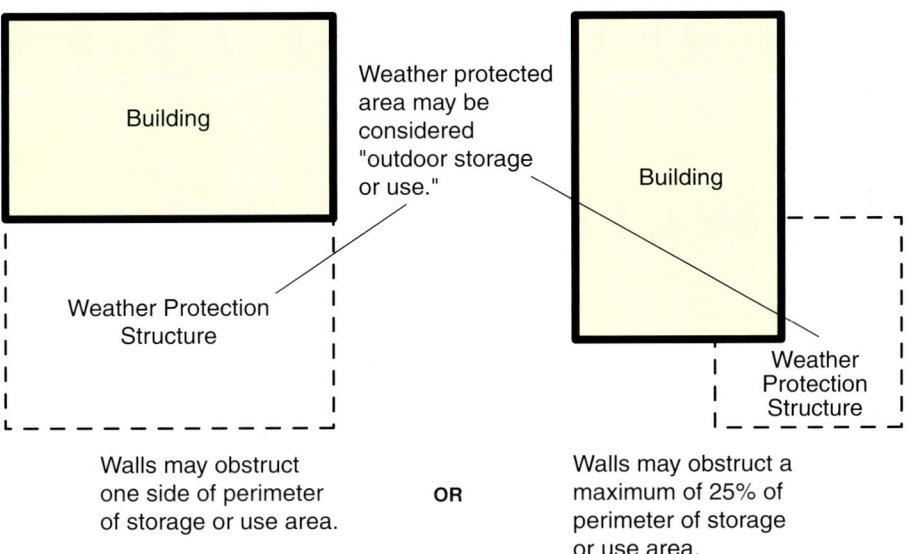

Outdoor Storage and Use of Hazardous Materials

3405.5

Alcohol-Based Hand Rubs Classified as Class I or II Liquids

Chapter 34 contains new requirements for alcohol-based hand sanitizer dispensers.

CHANGE TYPE. Addition

CHANGE SUMMARY. Section 3405.5 has been added to prescribe detailed requirements for installation of alcohol-based hand rub dispensers containing gelled flammable or combustible liquids. (F233-04/05)

2006 CODE: 3405.5 Alcohol-Based Hand Rubs Classified as Class I or II Liquids. The use of wall-mounted dispensers containing alcohol-based hand rubs classified as Class I or II liquids shall be in accordance with all of the following:

1. The maximum capacity of each dispenser shall be 68 ounces (2 L).
2. The minimum separation between dispensers shall be 48 inches (1219 mm).
3. The dispensers shall not be installed directly adjacent to, directly above or below an electrical receptacle, switch, appliance, device or other ignition source. The wall space between the dispenser and the floor shall remain clear and unobstructed.
4. Dispensers shall be mounted so that the bottom of the dispenser is a minimum of 42 inches (1067 mm) and a maximum of 48 inches (1219 mm) above the finished floor.
5. Dispensers shall not release their contents except when the dispenser is manually activated.
6. Storage and use of alcohol-based hand rubs shall be in accordance with the applicable provisions of Sections 3404 and 3405.
7. Dispensers installed in occupancies with carpeted floors shall only be allowed in smoke compartments or fire areas equipped throughout with an approved automatic sprinkler system in accordance with Section 903.3.1.1 or 903.3.1.2.

3405.5.1 Corridor Installations. Where wall-mounted dispensers containing alcohol-based hand rubs are installed in corridors, they shall be in accordance with all of the following:

1. Aerosol containers shall not be allowed in corridors.
2. The maximum capacity of each dispenser shall be 41 ounces (1.2 L).
3. The maximum quantity allowed in a corridor within a control area shall be 10 gallons (37.85 L).
4. The minimum corridor width shall be 72 inches (1829 mm).
5. Projections into a corridor shall be in accordance with Section 1003.3.3.

CHANGE SIGNIFICANCE. In 2003, the American Hospital Association (AHA) approached the ICC to express concern that some code officials were prohibiting the installation of alcohol-based hand rub

(ABHR) dispensers in hospital corridors. ABHR, typically packaged in pump bottles or bladders inserted into dispensers, are regarded by the health care industry as an important tool in fighting the spread of infectious diseases, and making the units very accessible to hospital staff is critical to encouraging their use. According to studies conducted by the Centers for Disease Control and Prevention (CDC), use of ABHR by staff increased by more than 20% when dispensers were installed immediately outside of patient/residence bedrooms or within suites of rooms.

ABHR has reportedly been in use for over 20 years in hospitals throughout Europe without a notable fire incident. A study of U.S. hospitals indicated widespread ABHR use as well. Approximately 95% of 840 U.S. hospitals that were surveyed indicated ongoing use of ABHR dispensers in rooms and/or corridors. With such widespread use, it is easy to see that fire code restrictions on ABHR in hospitals could have a significant impact on the health care industry.

In an effort to identify the most effective balance between fire safety concerns and the need to control the spread of infectious diseases, the ICC Board of Directors appointed an ad hoc committee to investigate these issues and to develop recommendations for code changes, as appropriate. The committee, which consisted of eight code officials and four industry representatives, produced a code change proposal that eventually became the new Section 3405.5.

One major concern in developing Section 3405.5 was the placing of ABHR dispensers in corridors of buildings, which also serve as the means of egress. To address this concern, additional limits were placed on ABHR dispensers located in corridors. It should be noted that all ABHR dispensers must meet the general requirements of Section 3405.5. Then, in addition, dispensers in corridors must meet the supplemental requirements in Section 3405.5.1. The additional limits for corridor installations include:

1. Prohibiting aerosol ABHR dispensers, which store ABHR under pressure,
2. Limiting the maximum capacity of any individual dispenser to 41 ounces,
3. Limiting the maximum aggregate quantity of ABHR in a corridor within an individual control area to 10 gallons (the 10-gallon limit is consistent with the requirements set forth in IFC Section 3404.3.4.4, which restrict the quantity of flammable and combustible liquids used for operations and maintenance to 10 gallons outside of a liquid storage cabinet),
4. Requiring a minimum corridor width of 72 inches (which largely restricts the use of ABHR in corridors to health care occupancies, since most other occupancies do not have corridors meeting these requirements), and
5. Requiring that the ABHR dispenser comply with limitations on projections into a corridor required for a means of egress under Chapter 10.

3405.5 continues

3405.5 continued

Regulations in Section 3405.5, including those in Section 3405.5.1, are based in part on the fire modeling that was sponsored by the American Society of Health Care Engineering (ASHE). Although the likelihood of igniting ABHR in a patient unit is considered remote, it is nevertheless important to understand the consequences of an ABHR fire before permitting the installation of dispensers in egress corridors. To evaluate the consequences of an ABHR fire, the models assumed complete rupture of a dispenser and ignition of the resulting liquid pool.

Modeling results indicated that both ethyl and isopropyl alcohol-based ABHR products can be used safely when dispenser sizes are limited to 2 liters in suites and 1.2 liters in minimum 6-ft-wide corridors. Horizontal spacing of 48 inches or more between dispensers was necessary to prevent ignition of fuel targets that were assumed in the models, so this limitation has been included in Section 3405.5. In addition, analysis of scenarios involving carpeted floors suggested that combustion of carpet could lead to visibility problems, so Section 3405.5 requires sprinkler protection where ABHR dispensers are installed in carpeted areas.

It should be noted that, in addition to the new Section 3405.5, a new Exception 10 to Section 2701.1 has been added to fulfill the intent of this code change by exempting ABHR from maximum allowable quantity limits and other controls in Chapter 27 that would ordinarily apply. For more information on the change to Section 2701.1, see the discussion of that section herein.

In addition, individuals responsible for applying or enforcing the IFC in health care occupancies will want to note that the new Section 3405.5 is generally consistent with regulations set forth in NFPA 101, Life Safety Code, another code that is often referenced in the regulation of hospitals and other health care occupancies. However, there are some differences between the IFC and NFPA 101 provisions. For example, the 10-gallon limit for dispensers in a single control area in the IFC parallels a 10-gallon limit per smoke compartment in NFPA 101. Control areas and smoke compartments are not necessarily constructed in an equivalent manner.

Finally, it should be noted that although the AHA initiated this revision on the basis of concerns related to health care occupancies, the new Section 3405.5 is not occupancy-specific. The provisions that were developed were deemed suitable for application in any occupancy.

CHANGE TYPE.   Modification

CHANGE SUMMARY.   Requirements applicable to separating flammable gases from exposures have been significantly expanded and now parallel the provisions applicable to other hazardous materials classification categories. **(F238-04/05)**

2006 CODE: 3504.2.1 ~~Outdoor Storage Areas~~ Distance Limitation to Exposures.   Outdoor storage ~~areas for~~ or use of flammable compressed gases shall be located from a lot line, public street, public alley, public way, or building not associated with the manufacture or distribution of such gases in accordance with Table 3504.2.1.

3504.2.1
Distance Limit to Exposures for Flammable Gases

TABLE 3504.2.1   Flammable Gases—Distance from Storage to Exposures

Maximum Amount per Storage Area (cubic feet)	Minimum Distance Between Storage Areas (feet)	Minimum Distance to Lot Lines of Property that Can be Built Upon (feet)[a]	Minimum Distance to Public Streets, Public Alleys or Public Ways (feet)[a]	Minimum Distance to Buildings on the Same Property		
				Nonrated Construction or Openings Within 25 Feet	2-Hour Construction and No Openings Within 25 Feet	4-Hour Construction and No Openings Within 25 Feet
0–4225	5	5	5	5	0	0
4226–21,125	10	10	10	10	5	0
21,126–50,700	10	15	15	20	5	0
50,701–84,500	10	20	20	20	5	0
84,501 or greater	20	25	25	20	5	0

For SI: 1 foot = 304.8 mm, 1 cubic foot = $0.02832 m^3$.

a. The minimum required distances shall not apply when fire barriers without openings or penetrations having a minimum fire-resistance rating of 2 hours interrupt the line of sight between the storage and the exposure. The configuration of the fire barrier shall be designed to allow natural ventilation to prevent the accumulation of hazardous gas concentrations.

CHANGE SIGNIFICANCE.   Section 3504.2.1 and Table 3504.2.1 have been significantly revised to expand the provisions governing the separation of flammable gases from exposures. It should be noted that the intent of Chapter 35 is to have Table 3504.2.1 govern both storage and use conditions. This is stated in Section 3504.2.1, which falls under Section 3504 (storage), but inclusion of a reference for use conditions in Section 3505 (use) and correction of the title to Table 3504.2.1 were inadvertently overlooked when the code was changed.

The revised provisions in Section 3504.2.1 accommodate the need by manufacturers and packagers of flammable compressed gases to place empty or partially full containers in staging areas outside of buildings used for manufacturing, packaging, and/or distribution of the same gases. In such cases, the proximity of containers immediately outside the building does not impact the level of hazard associated with the building or its contents. Similar allowances have been permitted in Section 3704.3.2.1.1 for toxic or highly toxic gases,

3504.2.1 continues

3504.2.1 continued   Table 4004.2.2 for oxidizing gases, and Section 4304.2.3 for unstable (reactive) materials.

Table 3504.2.1 has been revised by giving due consideration to the provisions for liquefied petroleum gases in Chapter 38, pyrophoric gases in Table 4104.2.1, and flammable solids in Chapter 36. The specified distances required by Table 3504.2.1 in the 2003 edition for separation between storage areas, lot lines, public streets, etc., have been maintained, and new distances have been added to specify required separations between the storage area and buildings on the same property.

The distances specified for separations from buildings were based in part on distances applicable to pyrophoric gases, with modifications made to account for the quantity ranges indicated in the 2003 version of Table 3504.2.1. A maximum required building-separation distance of 20 feet was selected to parallel maximum distances specified by the IFC for other materials, such as flammable solids (Section 3604.2.1), and the permissible use of a protective structure in lieu of separation distances, as allowed by Table 3504.2.1, Footnote "a," was based on provisions for flammable solids in Section 3604.2.1.

Outdoor storage and use areas for flammable gases, such as this one, must be properly separated from exposures.

CHANGE TYPE.   Modification

CHANGE SUMMARY.   Application of the code with respect to re-quiring gas detection for toxic gases has been clarified. **(F184-03/04)**

2006 CODE:

3702.1 PHYSIOLOGICAL WARNING THRESHOLD LEVEL.   <u>A con-centration of air-borne contaminants, normally expressed in parts per million (ppm) or milligrams per cubic meter (mg/m³), that represents the concentration at which persons can sense the presence of the con-taminant due to odor, irritation, or other quick-acting physiological re-sponses. When used in conjunction with the permissible exposure limit (PEL), the physiological warning threshold levels are those con-sistent with the classification system used to establish the PEL. See the definition of "Permissible exposure limit (PEL)" in Section 2702.</u>

3704.2.2.10 (IBC 908.3) Gas Detection System.   A gas detection system shall be provided to detect the presence of gas at or below the permissible exposure limit (PEL) or ceiling limit of the gas for which detection is provided. The system shall be capable of monitoring the discharge from the treatment system at or below one-half the IDLH limit.

> **Exception:** A gas detection system is not required for toxic gases when the physiological warning ~~properties~~ <u>threshold level</u> for the gas ~~are~~ <u>is</u> at a level below the accepted PEL for the gas.

1803.13 (IBC 415.9.7) Continuous Gas Detection Systems.   A continuous gas detection system shall be provided for HPM gases when the physiological warning ~~properties~~ <u>threshold level</u> of the gas ~~are~~ <u>is</u> at a higher level than the accepted permissible exposure limit (PEL) for the gas and for flammable gases in accordance with this section.

CHANGE SIGNIFICANCE.   To clarify the intent of the code, the term "physiological warning properties" has been replaced by the term "physiological warning threshold level," and a definition has been provided for the new term. The term "physiological warning proper-ties" caused problems because it was neither intuitive nor defined.

From a practical standpoint, the intent of the code in referencing physiological warning properties was to identify a concentration of a contaminant that would be detectable by an average individual on the basis of body warning signals such as odor or irritating effects (sting-ing sensations, coughing, scratchy feeling in the throat, running of the eyes or nose, or similar signals). The new term, "physiological warn-ing threshold level," more clearly conveys this intent, both in the term itself and with the newly provided definition.

The new definition also addresses a problem in identifying suit-able data specifying concentrations at which a release can be detected

3702.1, 3704.2.2.10 and 1803.13

Physiological Warning Thresholds

Gas detection is required for storage and use of some toxic gases.

3702.1, 3704.2.2.10 and 1803.13 continues

3702.1, 3704.2.2.10 and 1803.13
continued

by physiological response versus permissible exposure limits (PEL). Selecting appropriate data is important because the need for a gas detection system under Chapters 18 and 37 may be determined by the selected data.

In some cases, code officials have identified high odor thresholds, exceeding the PEL, for gases such as chlorine or ammonia to justify requiring gas detection for these gases. Many organizations publish data on gas properties, and this information may vary widely from one source to another based on test conditions, test methods, population exposed, and interpretation of results. It is not the intent of the code to permit "shopping" various data sources in order to make a case for more restrictive or more liberal code application with respect to requiring gas detection, and the new terminology clarifies this through the new definition.

The new definition links the determination of the physiological warning threshold level to the data used to determine the PEL (for example, PELs established by Title 29 of the Code of Federal Regulations [29CFR]).

For some hazardous materials, data developed by the American Conference of Governmental Industrial Hygienists (ACGIH) prescribing threshold limit values (TLVs) can be used in determining PELs (see Section 2702.1, definition of "permissible exposure limit"). To substantiate the TLVs (PELs), the ACGIH publishes the *Documentation of the Threshold Limit Values* and *Biological Exposure Indices,* which provide source data used in the establishment of TLVs. Included in these source data are perception thresholds, and these thresholds should be used as a basis for identifying physiological warning threshold levels when ACGIH TLVs (PELs) are used.

When other PEL data sources are utilized, any data on physiological warning properties published by the PEL source organization should be utilized.

By following the foregoing guidelines, as set forth in the definition of "physiological warning threshold level," proper application of the code with respect to requiring gas detection will be achieved.

CHANGE TYPE.   Modification

CHANGE SUMMARY.   The maximum container size for toxic gases in use permitted to apply the detect-shutoff option to treatment systems has been reduced. **(F184-03/04)**

2006 CODE: **3704.2.2.7 Treatment Systems.**   The exhaust ventilation from gas cabinets, exhausted enclosures, gas rooms, and local exhaust systems required in Sections 3704.2.2.4 and 3704.2.2.5 shall be directed to a treatment system. The treatment system shall be utilized to handle the accidental release of gas and to process exhaust ventilation. The treatment system shall be designed in accordance with Sections 3704.2.2.7.1 through 3704.2.2.7.5 and Section 510 of the *International Mechanical Code.*

3704.2.2.7 continues

3704.2.2.7

Treatment Systems for Toxic Gases

Use of the detection-shutoff option in lieu of a ventilation treatment system for toxic gases is now limited to containers having a maximum water capacity of 1700 pounds, such as these ton containers of chlorine. (Photo courtesy of Halogen Valve Systems, Inc., Irvine, CA.)

3704.2.2.7 continued

1. (No change to current text)

2. Toxic gases—use. Treatment systems are not required for toxic gases supplied by cylinders or portable tanks not exceeding ~~660 gallons (2498 L) liquid capacity~~ 1700 pounds (772 kg) water capacity when the following are provided:

    2.1. A gas detection system with a sensing interval not exceeding 5 minutes.

    2.2. An approved automatic-closing fail-safe valve located immediately adjacent to cylinder or portable tank valves. The fail-safe valve shall close when gas is detected at the PEL by a gas detection system monitoring the exhaust system at the point of discharge from the gas cabinet, exhausted enclosure, ventilated enclosure, or gas room. The gas detection system shall comply with Section 3704.2.2.10.

CHANGE SIGNIFICANCE. The maximum container size for "in use" toxic gases permitted to apply the detect-shutoff option to treatment systems has been reduced. The revised provisions are now limited to accommodate containers no larger than ton containers of chlorine. When the term "portable tank" was previously added to this section, it was intended to encompass ton containers of chlorine. However, because the term "portable tank" had been associated in some legacy codes with tanks having a 660-gallon liquid capacity, this inappropriately became the upper bound for portable tanks governed by this section.

Subsequent research was unable to identify a basis for permitting an upper limit on vessel size that goes beyond the capacity of a ton container. Accordingly, Section 3704.2.2.7 was revised to reflect the ton container limit, which is expressed as 1700 pounds water capacity.

The 1700-pound water capacity limit was derived as follows. A ton container typically holds about 1600 pounds of water, and a filling density of approximately 125% of the water capacity is permitted for chlorine (1600 ×1.25 = 2000 pounds). The resulting weight of product in a filled container is 1 ton for chlorine. The recommended water capacity of 1700 pounds vs. 1600 pounds is intended to accommodate manufacturing variations that occur from one container to the next, but this does not affect the ultimate gas capacity of a filled container, which is limited to 2000 pounds regardless of the variation in water capacity.

CHANGE TYPE.   Modification

CHANGE SUMMARY.   Most portable LP-gas containers are now required by the IFC to be equipped with overfill prevention devices prior to being refilled. **(F240-04/05)**

2006 CODE: 3806.2 Overfilling. ~~Liquefied petroleum~~ LP-gas containers shall not be filled or maintained with LP-gas in excess of either the volume determined using the fixed liquid-level gauge installed by the manufacturer, or the weight determined by the required percentage of the water capacity marked on the container. <u>Portable containers shall not be refilled unless equipped with an overfilling prevention device (OPD) when required by Section 5.7.6 of NFPA 58.</u>

CHANGE SIGNIFICANCE.   Portable LP-gas containers without overfill prevention devices (OPDs) are no longer permitted by the IFC to be refilled. This change improves correlation between the IFC and NFPA 58, which is the industry standard governing storage and use of liquefied petroleum gas (LP-gas) referenced by the IFC in Section 3801.1. An OPD, which is part of the valve assembly on an LP-gas container, serves as a backup means of overfill prevention. The primary method of avoiding overfills is limiting the amount of LP-gas in a container during filling on the basis of the weight or volume of gas.

Until relatively recently, a common method used to determine whether an LP-gas container had been filled to capacity was waiting for liquid to squirt from a small vent opening in the cylinder valve. A major problem with this method is that containers filled to capacity on a cold day might overpressurize and vent gas if later exposed to in-

3806.2 continues

3806.2

Overfill Prevention for LP-Gas Containers

Example of an overfill prevention device. Note that valve handles for overfill prevention devices are typically uniquely shaped as a triangle to help identify cylinders with this equipment.

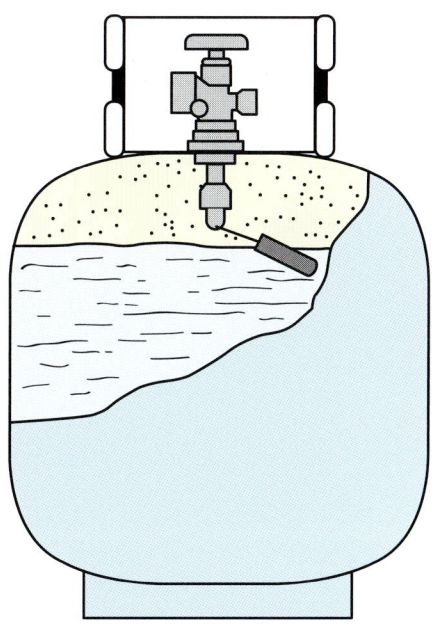

Illustration of an installed overfill prevention device. Note that the lever at the bottom of this particular device floats upward as the liquid level rises and ultimately seals the valve to prevent overfilling of the vessel.

The cylinder exchange industry has replaced millions of old-style cylinder valves with overfill prevention devices to comply with regulations mandating this equipment.

3806.2 continued creased temperatures. OPDs utilize a float valve to block the fill connection once a safe fill level has been reached, preventing such events.

NFPA 58 has required OPDs on most LP-gas containers in the 4- to 40-pound range since 2002, and most containers without OPDs have not been permitted by NFPA 58 to be refilled since that time. Thus, the change to Section 3806.2 is not entirely new. It simply coordinates the IFC with its reference standard.

CHANGE TYPE.   Modification

CHANGE SUMMARY.   The requirements for locating LP-gas containers awaiting use or resale have been updated to better correlate with NFPA 58, permitting cylinder exchange cabinets to be located in closer proximity to exits and building openings in some cases. **(F241-04/05)**

3809.12

LP-Gas Storage Outside of Buildings

2006 CODE: **3809.12 Location of Storage Outside of Buildings.** Storage outside of buildings, for containers awaiting use, resale, or part of a cylinder exchange program shall be located ~~not less than 20 feet (6096 mm) from openings into buildings, 20 feet (6096 mm) from any motor fuel dispenser and 10 feet (3048 mm) from any combustible material and~~ in accordance with Table 3809.12.

~~**TABLE 3809.12**   **Location of Containers Awaiting Use or Resale Stored Outside of Buildings**~~

~~Quantity of LP-Gas Stored~~	~~Distances to a Building or Group of Buildings, Public Way or Lot Line of Property that Can be Built Upon (feet)~~
~~500 pounds or less~~	~~0~~
~~501 to 2500 pounds~~	~~10ᵃ~~
~~2501 to 6000 pounds~~	~~15~~
~~6001 to 10,000 pounds~~	~~20~~
~~Over 10,000 pounds~~	~~25~~

~~SI — For SI: 1 Foot = 304.8 mm, 1 pound = 0.454 kg~~

~~a.   Containers are allowed to be located a lesser distance~~

TABLE 3809.12   **Separation of Containers Awaiting Use, Resale, or Exchange Stored Outside of Buildings from Exposures**

Quantity of LP-Gas Stored (pounds)	Minimum Separation Distance from Stored Cylinders to: (feet)						
	Nearest Important Building or Group of Buildings or Line of Adjoining Property that Can Be Built Upon	Line of Adjoining Property Occupied by Schools, Churches, Hospitals, Athletic Fields, or Other Points of Public Gathering; Busy Thoroughfares; or Sidewalks	LP-Gas Dispensing Station	Doorway or Opening to a Building with Two or More Means of Egress	Doorway or Opening to a Building with One Means of Egress	Combustible Materials	Motor Vehicle Fuel Dispenser
720 or less	0	0	5	5	10	10	20
721–2500	0	10	10	5	10	10	20
2501–6000	10	10	10	10	10	10	20
6001–10,000	20	20	20	20	20	10	20
Over 10,000	25	25	25	25	25	10	20

3809.12 continues

3809.12 continued   ~~**3809.14 Separation from Means of Egress for Containers Located Outside of Buildings.** Containers located outside of buildings shall not be located within 20 feet (6096 mm) of any exit access doors, exits, stairways, or in areas normally used, or intended to be used, as a means of egress.~~

CHANGE SIGNIFICANCE. Revisions to Section 3809.12 and Table 3809.12 and deletion of Section 3809.14 significantly improve the clarity of the code by replacing the existing mix of multiple text sections and a table with a single table that summarizes all relevant requirements for outdoor storage of LP gas containers awaiting use, resale, or exchange. The revisions also update the requirements to improve correlation with the 2004 edition of NFPA 58. Provisions in prior editions of the IFC were based on the 1983 and 1998 editions of NFPA 58, which are now well out of date.

Coordination of Section 3809.12 with the 2004 edition of NFPA 58 resulted in the following technical changes to IFC Table 3809.12:

1. **"Quantity of LP Gas Stored" column and separations to buildings and property lines:** The LP gas quantity ranges for the first two rows of Table 3809.12 have been modified to correlate with the ranges in NFPA 58–2004, Table 8.4.1.2. The old Footnote "a" provided for no building or property separations in the 501–2500 pound range, and this is consistent with the

Minimum separation distances between LP gas cylinder exchange cabinets and exposures have been revised.

current edition of NFPA 58, as reflected in the new table. The specified separation to buildings and property lines for the 2501–6000 pound range has been changed from 15 feet to 10 feet to correlate with the current edition of NFPA 58. There is no justification in the record to substantiate why the 1988 UFC (which served as the source for prior IFC provisions) specified a 15-foot distance versus the 10-foot distance in NFPA 58. The 10-foot distance has been in NFPA 58 for at least 30 years and served as the regulation in jurisdictions using the BOCA National and Standard Fire Prevention Code, with no indication of inadequacy.

2. **Separations to public gathering places, thoroughfares, and sidewalks:** Provisions in the new Table 3809.12 were derived directly from NFPA 58 and enhance the level of safety provided by the IFC because of their improved visibility in the consolidated table. Regulations based on thoroughfares and sidewalks versus public ways correlate with terminology in NFPA 58. Separation distances for the 721–2500 pound range are now specified as 10 feet in all cases, with no allowance for reductions that were permitted by the prior Footnote "a." Separation distances for the 2501–6000 pound range have been reduced from 15 feet to 10 feet, as explained in previous Item 1.

3. **Separations to LP gas dispensing stations:** The provisions in the new Table 3809.12 were derived directly from NFPA 58 and enhance the level of safety provided by the IFC because of their improved visibility in the consolidated table.

4. **Separation to exitways and building openings:** The 20-foot separation distance requirements applicable to exit elements and building openings in prior editions of the IFC have been reduced as follows:

   a. 10 feet between LP gas and any doorway or opening to a building with a single means of egress, for installations not exceeding 6000 pounds of LP gas

   b. 5 feet between LP gas and any doorway or opening to a building with two or more means of egress, for installations not exceeding 2500 pounds of LP gas

   c. 10 feet between LP gas and any doorway or opening to a building with two or more means of egress, for installations of 2501 to 6000 pounds of LP gas.

The 20-foot separation distance requirements in prior editions of the IFC originated with Proposals F872–99 and F873–99. The stated intent of those proposals was correlation with NFPA 58, and in fact, they did generally correlate with the 1998 edition of NFPA 58. However, the 20-foot separation provisions in NFPA 58–1998 were very short-lived.

The 20-foot distances were added to NFPA 58 as a result of a floor action at the Fall 1997 NFPA final action hearing, which passed by a

3809.12 continues

3809.12 continued vote of 59 to 54. The NFPA Standards Council upheld the floor action when they issued NFPA 58–1998, but they then directed that a special task group be convened to resolve differences of opinion between interested parties for the purpose of producing a Tentative Interim Amendment (TIA) to the 1998 edition. As directed, the Task Group, which included both fire service and industry interests, developed a compromise position that was forwarded to the Standards Council, and the compromise was published as a TIA, which became effective on October 28, 1998.

Meanwhile, the closing date for proposals to the 1999 code cycle for the IFC had passed, and Proposals F872–99 and F873–99 were subsequently approved, incorporating the 20-foot distances published in NFPA 58–1998 into the IFC instead of the revised TIA distances. The TIA required a minimum 10-foot separation to doorways and building openings for buildings with one means of egress and a minimum 5-foot separation to doorways and building openings for buildings with two or more means of egress. Following publication of the TIA, the TIA provisions were adopted for inclusion in the 2001 edition of NFPA 58, and they were retained without change in the 2004 edition, indicating that the controversy involving the 20-foot versus lesser distance requirements had been resolved.

Beyond the goal of correlation, there are valid technical and statistical reasons for substituting the updated NFPA 58 distances to doorways and building openings for the outdated IFC requirements. These include the following:

i. Overfill prevention devices: Since 2002, NFPA 58 has not permitted the filling of most cylinders for vapor service in the 4- to 40-pound range unless the cylinder is equipped with an overfill prevention device (OPD). Accordingly, all cylinders in a cylinder exchange program must have been originally equipped with or be retrofitted with an OPD. Prior to the introduction of OPDs, overfilled cylinders may have initially held their charge but later leaked when cylinder pressure increased because of high ambient temperature exposures. Such leakage scenarios were a significant source of concern when the 20-foot doorway/opening separation distances were initially justified in the IFC, but today, cylinder exchange operators report that cylinder leaks due to overfill have declined to almost zero as a result of the use of OPDs and other overfill prevention methodologies, such as weight- and volume-based filling.

ii. Release modeling: Release modeling conducted by the Austin Fire Department in the 1990s and quoted as justification for separation distances in Proposal F872–99 assumed a leaking cylinder due to overfill and then a relief valve that stuck in the open position, releasing the entire contents of a single cylinder. Such a scenario is no longer plausible with the mandatory use of OPDs.

iii. Leak-resistant cylinder valves: Although not specifically mandated, most new grill cylinder valves are designed such that they will flow gas only if a "positive seal" is maintained between the mating valve on the cylinder and an appliance connection. Otherwise, an internal check valve remains closed and prevents the escape of gas, even if the cylinder valve is not shut.

iv. Loss statistics: A survey of companies responsible for over 100,000 cylinder exchange cabinets accessible to the public (the vast majority of those that are installed) covering the past 10 years revealed that that there is no record of fire involving the release of propane from a cylinder in a cylinder exchange cabinet. In addition, there have been no reports of gas-release incidents resulting from a vehicular collision with a cylinder exchange cabinet.

v. Scenario analysis: If a fire occurs inside of a building, the presence of an LP gas exchange cabinet outside is not an egress or life-safety concern. Individuals leaving the building can simply exit past the cabinet. If fire occurs outside, the fire would have to achieve a size that is adequate to raise the temperature of stored LP gas to the 150 degrees-Fahrenheit range before a standard container relief vent would be expected to operate. Given the minimum required 10 foot separation to combustibles required by Table 3809.12, it is reasonable to assume that building occupants would have time to evacuate a single-exit building before an exposure fire would cause container venting. Scenarios involving a release of gas from an overfilled container and subsequent migration to an ignition source are no longer regarded as plausible with the mandatory use of OPDs and the likely use of leak-resistant container valves as described previously.

It should be noted that one deliberate difference between the IFC and NFPA 58 pertains to the minimum required clearance to combustibles. The 10-foot minimum clearance to combustibles, which was included in prior editions of the IFC, has been retained even though NFPA 58 uses a performance requirement in lieu of specifying a prescriptive distance. Continuing this inconsistency was a specific desire of the fire service based on the need for the IFC to include a clearly stated, enforceable requirement.

PART 6

Reference Standards

Chapter 45

■ **Chapter 45** Reference Standards No changes
addressed

Chapter 45 contains all of the standards adopted by reference in the *International Fire Code.* These standards provide testing, design and installation criteria to support IFC requirements. Most are written by standards-writing organizations or various industry trade associations. Standards referenced in Chapter 45 are automatically updated to their most current editons availale at the time just prior to publication of each new IFC edition. ■

PART 7

Appendices
Appendices A Through G

- **Appendix A** Board of Appeals No changes addressed
- **Appendix B** Fire-Flow Requirements for Buildings
- **Appendix C** Fire Hydrant Locations and Distribution No changes addressed
- **Appendix D** Fire Apparatus Access Roads No changes addressed
- **Appendix E** Hazard Categories No changes addressed
- **Appendix F** Hazard Ratings No changes addressed
- **Appendix G** Cryogenic Fluids—Weight and Volume Equivalents No changes addressed

Appendices A through G contain requirements for establishing Board of Appeals, fire department access, fire flow requirements for buildings and locations for fire hydrants. The remaining appendices provide information concerning the classification of hazardous materials and the equivalent weights and volumes of the most commonly found cryogenic fluids. Appendices A through D do not have the force of law unless they are adopted by a jurisdiction. Appendices E through G are intended as information sources that supplement the requirements of the *International Fire Code* and are not intended for adoption. ■

APPENDIX B105.2

Fire-Flow Requirements for Buildings Other Than One- and Two-Family Dwellings

Appendix B105.2

Fire-Flow Requirements for Buildings Other Than One- and Two-Family Dwellings

CHANGE TYPE.   Modification

CHANGE SUMMARY.   A change to the 2003 edition has been reversed, reinstating the maximum 75% permissible reduction for sprinklers, regardless of a building's type of construction, occupancy, or NFPA 13 versus 13R protection. **(F244-04/05)**

2006 CODE: B105.2 Buildings Other Than One- and Two-Family Dwellings.   The minimum fire flow and flow duration for buildings other than one- and two-family dwellings shall be as specified in Table B105.1.

> **Exception:** A reduction in required fire flow of up to ~~50~~ 75 percent, as approved, is allowed when the building is provided with an approved automatic sprinkler system installed in accordance with Section 903.3.1.1 or 903.3.1.2. ~~of the *International Fire Code.* Where buildings are also of Type I or II construction and are a light-hazard occupancy as defined by NFPA 13, the re-~~

When a building is equipped with an automatic sprinkler system complying with either NFPA 13 or 13R, up to a 75% reduction in required fire flow is permitted. (Photo courtesy of Protection Development, Inc.)

~~duction may be up to 75 percent.~~ The resulting fire flow shall not be less than 1500 gallons per minute (5678 L/min) for the prescribed duration as specified in Table B 105.1.

CHANGE SIGNIFICANCE.   Section B105.2 has reverted to provisions that existed in the 2000 edition of the IFC. A revision to the 2003 edition, which is no longer in effect, reduced the maximum permissible fire-flow reduction under this section to 50% unless a building was built in accordance with the requirements for Type I or Type II construction, housed a light hazard occupancy, and was protected by a sprinkler system complying with NFPA 13.

The change to the 2003 edition was reversed because several errors in the substantiation for Proposal F126–01, which served as the basis for the change, were later identified. In addition, the cost impact of the revision to the 2003 edition was not well known when the revision was initially considered by the ICC membership. Subsequent analysis has revealed that the cost impact of the 2003 revision was significant and not warranted on the basis of the supporting justification.

For example, assuming the maximum permissible reduction, the required fire flow for a typical retail store with Type II-B construction and an area of 100,000 square feet grew from 1690 gpm to 3375 gpm as a result of the 2003 edition change. Given that many water systems are not capable of providing more than 3000 gpm in retail areas, costs of construction were increased significantly with the 2003 requirements because additional firewalls or increased construction types became necessary to satisfy the code. Alternately, on-site water storage tanks and fire pumps to supply fire hydrants could be used, but all of the foregoing options are very costly, and such expense could not be justified by a cost-benefit analysis.

It should be noted that the permissible reduction in fire flow remains a subjective decision of the fire code official. Simply because the code now permits a reduction of up to 75% for sprinklered buildings, there is no guarantee that the maximum reduction is suitable or acceptable for every case. The change in the 2006 edition simply returns full discretion to the local fire code official with regard to the percentage reduction to be granted.

It should also be noted that an additional change in the 2003 edition to Table B105.1, Footnote "a," permitted a 25% reduction in the tabular fire flow values for Group R occupancies, which essentially negated the impact of the 2003 edition change to Section B105.2 with respect to Group R. Because this footnote remains in the 2006 edition, even though the change to Section B105.2 has been reversed, the required fire flow for Group R will actually be less than was required by the 2000 edition if the maximum fire-flow reductions are permitted. This was not an intentional change, and users of the code are encouraged to permit no more than a 75% total reduction from the values in Table B105.1 for Group R occupancies to maintain consistency with the intent of the code. A proposal to fix this issue has been processed for the 2009 IFC.

Index

A

Aboveground tanks, minimum separation
  requirements for, 175, 191–192
Administration, 1–7
  change of use or occupancy, 1, 2–3
  missed violations-approvals, 1, 6–7
  permit amounts for hazardous materials, 1, 4–5
Aerosols, 202, 203, 218–222
Aisle accessways in Group M occupancies, 43,
  160–161
Alarm communications systems. *See* Voice/alarm
  communication systems
Alarm notification appliances in employee work
  areas, 42, 94
Alcohol-based hand rubs classified as Class I or II
  liquids, 203, 258–260
Appendices, 275–277
  fire-flow requirements for buildings, 275, 276–277
  one- and two-family dwellings, 275, 276–277
Appliances, alarm notification, 42, 94
Assembly Group A, definition for, 1, 12–13
Automatic fire sprinkler installation requirements,
  42, 76–77
Automatic sprinkler systems
  family dwellings and townhouses, 42, 72–73
  Group A-2, 42, 74–75

B

Balconies and decks, 42, 76–77
Basement storage, 203, 250–251
Battery systems, stationary, 41, 54–58
Boatyards, standpipe systems for marinas and, 42,
  82–83
Building and site requirements, 41–173
  building services and systems, 41
  fire protection systems, 41
  interior finish, decorative materials, andfurniture,
  41
  means of egress, 41
Building services and systems, 41, 44–58
  emergency control boxes for refrigeration
  systems, 41, 48–49
  emergency pressure control systems for
  refrigeration systems, 41, 50–53

portable electric space heaters, 41, 44–45
refrigerant leak detectors, 41, 46
remote controls for refrigeration systems, 41, 47
stationary battery systems, 41, 54–58
Buildings
  fire-flow requirements for, 275, 276–277
  Group R-2 college and university, 23, 38–39
  LP-gas storage outside of, 203, 269–273
  storage of fueled equipment in, 23, 28–29
  unsafe, 23, 26–27
Business Group B, definition for, 1, 14

C

Cabinets, storage, 203, 249
Ceiling height, minimum, 42, 111
Class I standpipe hose connections at horizontal
  exits, 42, 84–85
Class K fire extinguishers for deep-fat fryers, 42,
  80–81
College and university buildings, Group R-2, 23,
  38–39
Combustible fibers
  control area for, 202, 207–208
  definitions of, 203, 223–224
Combustible liquids, flammable and, 202, 203,
  242–260
  alcohol-based hand rubs classified as Class I or II
  liquids, 203, 258–260
  basement storage, 203, 250–251
  definition of liquid storage warehouse, 203,
  242–243
  design, construction, and capacity of containers,
  203, 246–248
  design, construction, and capacity of portable
  tanks, 203, 246–248
  number of storage cabinets, 203, 249
  overfill prevention, 203, 244–245
  quantity limits for liquid storage warehouses,
  203, 255
  storage of unsaturated polyester resin, 203, 252–254
  weather protection, 203, 256–257
Combustible storage, high-piled, 174, 175, 199–200
  plastic pallets and shelves in rack storage, 175,
  199–200

Communication systems
  for areas of refuge, 43, 173
  emergency voice/alarm, 42, 92–93
Compressed gases, 202, 203, 225–230
  containers, cylinders, and tanks, 203, 225–226
  pressure relief for, 203, 225–226
  vaults for, 203, 227–230
Containers
  compressed gas, 203, 225–226
  design, construction, and capacity of, 203,
    246–248
  overfill prevention for LP-gas, 203, 267–268
Control areas
  design and number of, 203, 214–215
  fire-resistance rating requirements for, 203,
    216–217
Control boxes for refrigeration systems, emergency,
    41, 48–49
Control systems, pressure, 41, 50–53
Controls and valves, definitions for emergency
    shutoff, 1, 8–9
Cooking appliances, portable fire extinguishers for
    solid-fuel, 42, 78–79
Curtains, draft, 42, 101–103
Curved stairways, 43, 136–137
Cylinders, compressed gas, 203, 225–226

D

Day care uses
  occupant load determination for, 42, 116–117
  single means of egress from, 43, 162–163
Decks, balconies and, 42, 76–77
Decorative materials, definitions for, 1, 10
Decorative materials, interior finish, and furniture,
    41, 59–67
Deep-fat fryers, fire extinguishers for, 42, 80–81
Definitions, 1, 8–22
  for Assembly Group A, 1, 12–13
  for Business Group B, 1, 14
  for decorative materials, 1, 10
  for emergency shutoff controls and valves, 1, 8–9
  for fail safe, 1, 11
  for high-hazard Group H, exceptions, 1, 15–17
  for high-hazard Group H-3, 1, 18–19
  for Residential Group R, 1, 20–21
  for Storage Group S-1, 1, 22
Design occupant load, 42, 114–115
Detection systems, monitoring of fire alarm and,
    95–96
Detectors, refrigerant leak, 41, 46

Door swings in sleeping units, 43, 128–129
Door width in Group R-1 occupancies, minimum,
    43, 126–127
Doors
  residential exterior, 43, 130–131
  stairway, 43, 132–133
Dormitories, open flame ignition sources in Group
    R-2, 23, 24
Draft curtains for high-piled storage, smoke venting
    and, 42, 101–103
Drills, emergency evacuation, 23, 35–37
Dwellings, family, 275, 276–277

E

Egress
  from Group A occupancies, 43, 170
  means of, 42–43, 110–173
  platform lifts of accessible means of, 42, 122
  reliability—maintenance of the means of, 43, 171
  separation of three or more exits, 43, 164–165
  from smokeproof enclosures, 43, 168–169
  through adjoining tenant spaces, 43, 156–157
  through intervening spaces, 43, 153–155
Egress, means of, 41, 42, 43, 110–173
  aisle accessways in Group M occupancies, 43,
    160–161
  curved stairways, 43, 136–137
  definition of accessible means of egress, 42, 110
  determination of design occupant load, 42,
    114–115
  door swings in sleeping units, 43, 128–129
  edge protection at ramps, 43, 149–150
  egress from Group A occupancies, 43, 170
  egress from smokeproof enclosures, 43, 168–169
  egress separation of three or more exits, 43,
    164–165
  egress through adjoining tenant spaces, 43,
    156–157
  egress through intervening spaces, 43, 153–155
  enclosed usable space under stairways, 43,
    140–141
  exit signs—maintenance, 43, 172
  guard opening limitations for Group R-2
    occupancies, 43, 151–152
  handrails for stairways and ramps, 43, 142–144
  maximum occupant load permitted, 42, 118–119
  minimum ceiling height, 42, 111
  minimum door width in Group R-1 occupancies,
    43, 126–127
  minimum ramp length, 43, 147–148

occupant load determination for day care uses, 42, 116–117

occupant load determination for fixed seating, 42, 120–121

panic and fire exit hardware, 43, 134–135

path of egress travel in Group R-2 occupancies, 43, 158–159

platform lifts as accessible means of egress, 42, 122

projection limits on freestanding objects, 42, 112–113

protection at roof-hatch openings, 43, 145–146

reliability—maintenance of the means of egress, 43, 171

remote unlocking of stairway doors, 43, 132–133

required areas of refuge, 42, 123–125

single means of egress from day care uses, 43, 162–163

testing and maintenance—communications systems for areas of refuge, 43, 173

thresholds at residential exterior doors, 43, 130–131

unenclosed interior exit stairways, 43, 166–167

weather protection of exterior egress components, 43, 138–139

Egress components, weather protection of exterior, 43, 138–139

Egress from day care uses, single means of, 43, 162–163

Egress travel in Group R-2 occupancies, path of, 43, 158–159

Electric space heaters, portable, 41, 44–45

Electronic monitoring for portable fire extinguishers, 42, 86–87

Emergency control boxes for refrigeration systems, 41, 48–49

Emergency evacuation drills, frequency of, 23, 35–37

Emergency planning and preparedness, 30–39

fire safety and evacuation plans for Group B, 23, 30–31

fire safety and evacuation plans for Group R-2, 23, 32–34

frequency of emergency evacuation drills, 23, 35–37

for Group E occupancies, 23, 38–39

for Group R-2 college and university buildings, 23, 38–39

Emergency pressure control systems for refrigeration systems, 41, 50–53

Emergency shutoff controls and valves, definitions for, 1, 8–9

Emergency voice/alarm communication systems, 42, 92–93

Employee work areas, alarm notification appliances in, 42, 94

Equipment, refueling of powered industrial trucks and, 23, 25

ESFR sprinklers, smoke and heat vents for, 42, 97–98

Evacuation drills, emergency, 23, 35–37

Evacuation plans for Group B, fire safety and, 23, 30–31

Evacuation plans for Group R-2, fire safety and, 23, 32–34

Exit signs—maintenance, 43, 172

Exit stairways, unenclosed interior, 166–167

Exits, egress separation of three or more, 43, 164–165

Explosives
control area for, 202, 205–206
day boxes and operating buildings, 203, 238–239
definition of, 203, 231–232
separation distances for, 203, 233–237, 238–239

Explosives and fireworks, 202, 203, 231–241
day boxes and operating buildings, 203, 238–239
definition of explosives, 203, 231–232
permit application for fireworks, 203, 240
post-fireworks display inspection, 203, 241
separation distances for explosives, 203, 233–237, 238–239

Exterior doors, residential, 43, 130–131

Exterior egress components, weather protection of, 43, 138–139

F

Fabrication facilities, semiconductor, 174, 175, 185–188
hazardous production materials at a workstation, 175, 185–188

Facilities
motor fuel-dispensing, 189–198
semiconductor fabrication, 174, 175, 185–188

Facilities, motor fuel-dispensing, 174, 175, 189–198
equipment for hydrogen motor fuel, 175, 193–198
minimum separation requirements for aboveground tanks, 175, 191–192
warning signs, 175, 189–190

Fail safe, definition for, 1, 11

Family dwellings
automatic sprinkler systems, 42, 72–73
one- and two-, 275, 276–277

Fibers, combustible, 202, 203, 207–208, 223–224
Finishes, flammable, 174, 175, 176–184
Fire, general precautions against, 23–29
    open flame ignition sources in Group R-2
        dormitories, 23, 24
    placards for unsafe buildings, 23, 26–27
    refueling of powered industrial trucks and
        equipment, 23, 25
    storage of fueled equipment in buildings, 23,
        28–29
Fire alarm and detection systems, monitoring of, 42,
        95–96
Fire alarm boxes in Group R-2, manual, 42, 91
Fire alarm systems for Group I, 42, 88–90
Fire exit hardware, panic and, 43, 134–135
Fire extinguishers, electronic monitoring for
        portable, 42, 86–87
Fire extinguishers for deep-fat fryers, Class K, 42,
        80–81
Fire extinguishers for solid-fuel cooking appliances,
        portable, 42, 78–79
Fire protection components, recall of, 42, 70–71
Fire protection for special uses and occupancies,
        42, 104–109
Fire protection systems, 41, 42, 68–109
    alarm notification appliances in employee work
        areas, 42, 94
    automatic fire sprinkler installation requirements,
        42, 76–77
    automatic sprinkler systems, 42
    automatic sprinkler systems—family dwellings
        and townhouses, 72–73
    balconies and decks, 42, 76–77
    Class I standpipe hose connections at horizontal
        exits, 42, 84–85
    Class K fire extinguishers for deep-fat fryers, 42,
        80–81
    electronic monitoring for portable fire
        extinguishers, 42, 86–87
    emergency voice/alarm communication systems,
        42, 92–93
    fire alarm systems for Group I, 42, 88–90
    fire protection for special uses and occupancies,
        42, 104–109
    Group A-2—automatic sprinkler systems, 42,
        74–75
    manual fire alarm boxes in Group R-2, 42, 91
    monitoring of fire alarm and detection systems,
        42, 95–96
    portable fire extinguishers for solid-fuel cooking
        appliances, 42, 78–79
    recall of fire protection components, 42, 70–71

records for fire protection systems, 42, 68–69
    smoke and heat vents for ESFR sprinklers, 42,
        97–98
    smoke and heat vents for Group H, 42, 99–100
    smoke venting and draft curtains for high-piled
        storage, 42, 101–103
    standpipe systems for marinas and boatyards, 42,
        82–83
Fire safety and evacuation plans for Group B, 23,
        30–31
Fire safety and evacuation plans for Group R-2, 23,
        32–34
Fire sprinkler installation requirements, automatic,
        42, 76–77
Fire-flow requirements for buildings, 275, 276–277
Fire-resistance rating requirements for control areas,
        203, 216–217
Fireworks. *See also* Post-fireworks display
        inspection
    permit application for, 203, 240
Fireworks, explosives and, 202, 203, 231–241
    day boxes and operating buildings, 203, 238–239
    definition of explosives, 203, 231–232
    permit application for fireworks, 203, 240
    post-fireworks display inspection, 203, 241
    separation distances for explosives, 203, 233–237,
        238–239
Fixed seating, occupant load determination for, 42,
        120–121
Flammable and combustible liquids, 202, 203,
        242–260
    alcohol-based hand rubs classified as Class I or II
        liquids, 203, 258–260
    basement storage, 203, 250–251
    definition of liquid storage warehouse, 203,
        242–243
    design, construction, and capacity of containers,
        203, 246–248
    design, construction, and capacity of portable
        tanks, 203, 246–248
    number of storage cabinets, 203, 249
    overfill prevention, 203, 244–245
    quantity limits for liquid storage warehouses,
        203, 255
    storage of unsaturated polyester resin, 203,
        252–254
    weather protection, 203, 256–257
Flammable finishes, 174, 175, 176–184
    location of spray-finishing operations, 175,
        182–183
    ventilation termination point, 175, 184
Flammable gases, 202, 203, 261–262

Freestanding objects, projection limits on, 42, 112–113

Fryers, deep-fat, 42, 80–81

Fuel, dispensing operations and equipment for hydrogen motor, 175, 193–198

Fuel in fuel tanks, control area for, 203, 209–210

Fuel in piping systems, control area for, 203, 209–210

Fuel tanks, control area for fuel in, 203, 209–210

Fuel-dispensing facilities, motor, 174, 175, 189–198
    equipment for hydrogen motor fuel, 175, 193–198
    minimum separation requirements for aboveground tanks, 175, 191–192
    warning signs, 175, 189–190

Fueled equipment, storage of, 23, 28–29

Furniture, interior finish, and decorative materials, 41, 59–67

G

Garages, repair, 174, 175, 189–198
    equipment for hydrogen motor fuel, 175, 193–198
    minimum separation requirements for aboveground tanks, 175, 191–192
    warning signs, 175, 189–190

Gases
    compressed, 202, 203, 225–230
    flammable, 202, 203, 261–262
    liquefied petroleum, 202, 203, 267–273
    vaults for compressed, 203, 227–230

Gases, compressed
    containers, cylinders, and tanks, 203, 225–226
    pressure relief for, 203, 225–226
    vaults for compressed gases, 203, 227–230

Gases, flammable
    distance limit to exposures for, 202, 203, 261–262

Gases, liquefied petroleum (LP), 202
    LP-gas storage outside of buildings, 203, 269–273
    overfill prevention for LP-gas containers, 203, 267–268

Gases, toxic, 202, 203, 265–266

Group A occupancies, egress from, 43, 170

Group A-2—automatic sprinkler systems, 42, 74–75

Group B, fire safety and evacuation plans for, 23, 30–31

Group E occupancies, emergency planning and preparedness for, 23, 38–39

Group H. smoke and heat vents for, 42, 99–100

Group I, fire alarm systems for, 42, 88–90

Group M occupancies, aisle accessways in, 43, 160–161

Group R-1 occupancies, minimum door width in, 43, 126–127

Group R-2
    fire safety and evacuation plans for, 23, 32–34
    manual fire alarm boxes in, 42, 91

Group R-2 college and university buildings, emergency planning and preparedness for, 23, 38–39

Group R-2 dormitories, open flame ignition sources in, 23, 24

Group R-2 occupancies
    guard opening limitations for, 43, 151–152
    path of egress travel in, 43, 158–159

Guard opening limitations for Group R-2 occupancies, 43, 151–152

H

Hand rubs, alcohol-based, 203, 258–260

Handrails for stairways and ramps, 43, 142–144

Hardware, panic and fire exit, 43, 134–135

Hatch openings. *See* Roof-hatch openings

Hazardous material, 202–273
    aerosols, 202, 203, 218–222
    combustible fibers, 202, 203, 223–224
    compressed gases, 202, 203, 225–230
    control area for combustible fibers, 202, 207–208
    control area for explosives, 202, 205–206
    control area for fuel in fuel tanks, 203, 209–210
    control area for fuel in piping systems, 203, 209–210
    design and number of control areas, 203, 214–215
    explosives and fireworks, 202, 203, 231–241
    fire-resistance rating requirements for control areas, 203, 216–217
    flammable and combustible liquids, 202, 203, 242–260
    flammable gases, 202, 203, 261–262
    highly toxic and toxic materials, 202, 203, 263–266
    liquefied petroleum gases, 202, 203, 267–273
    scope, 202, 204
    testing of hazardous materials equipment, devices, and systems, 203, 211–213

Hazardous materials equipment, devices, and systems, 203, 211–213

Hazardous materials, permit amounts for, 1, 4–5

Hazardous production materials (HPMs) at a workstation, 175, 185–188

Heat vents, smoke and, 42, 97–98, 99–100

Heaters, portable electric space, 41, 44–45

High-hazard Group H, exceptions, definition for, 1, 15–17

High-hazard Group H-3, definition for, 1, 18–19

High-piled combustible storage, 174, 175, 199–200
  plastic pallets and shelves in rack storage, 175,
    199–200

High-piled storage, smoke venting and draft
    curtains for, 42, 101–103

Hose connections at horizontal exits, Class I
    standpipe, 42, 84–85

HPMs. *See* Hazardous production materials.

Hydrogen motor fuel, dispensing operations and
    equipment for, 175, 193–198

I

Ignition sources in Group R-2 dormitories, open
    flame, 23, 24

Interior exit stairways, unenclosed, 43, 166–167

Interior finish, decorative materials, and furniture,
    41, 59–67

Intervening spaces, egress through, 43, 153–155

L

Lifts, platform, 42, 122

Liquefied petroleum (LP) gases, 202, 203, 267–273
  LP-gas storage outside of buildings, 203,
    269–273
  overfill prevention for LP-gas containers, 203,
    267–268

Liquid storage warehouses
  definition of, 203, 242–243
  quantity limits for, 203, 255

Liquids, alcohol-based hand rubs classified as Class
    I or II, 203, 258–260

Liquids, flammable and combustible, 202, 203,
    242–260
  alcohol-based hand rubs classified as Class I or II
    liquids, 203, 258–260
  basement storage, 203, 250–251
  definition of liquid storage warehouse, 203,
    242–243
  design, construction, and capacity of containers,
    203, 246–248
  design, construction, and capacity of portable
    tanks, 203, 246–248
  number of storage cabinets, 203, 249
  overfill prevention, 203, 244–245
  quantity limits for liquid storage warehouses,
    203, 255
  storage of unsaturated polyester resin, 203,
    252–254
  weather protection, 203, 256–257

Loads
  design occupant, 42, 114–115
  occupant, 42, 114–115, 118–119, 120–121

LP. *See* Liquefied petroleum

M

Manual fire alarm boxes in Group R-2, 42, 91

Marinas and boatyards, standpipe systems for, 42,
    82–83

Material, hazardous, 202–273
  aerosols, 202, 203, 218–222
  combustible fibers, 202, 203, 223–224
  compressed gases, 202, 203, 225–230
  control area for combustible fibers, 202, 207–208
  control area for explosives, 202, 205–206
  control area for fuel in fuel tanks, 203, 209–210
  control area for fuel in piping systems, 203,
    209–210
  design and number of control areas, 203, 214–215
  equipment, devices, and systems, 203, 211–213
  explosives and fireworks, 202, 203, 231–241
  fire-resistance rating requirements for control
    areas, 203, 216–217
  flammable and combustible liquids, 202, 203,
    242–260
  flammable gases, 202, 203, 261–262
  highly toxic and toxic materials, 202, 203, 263–266
  liquefied petroleum gases, 202, 203, 267–273
  scope, 202, 204
  testing of, 203, 211–213

Materials
  decorative, 1, 10
  hazardous, 1, 4–5

Materials, toxic, 202, 203, 263–266
  physiological warning thresholds, 203, 263–264
  treatment systems for toxic gases, 265–266

Means of egress, 41, 42, 43, 110–173

Missed violations-approvals, 1, 6–7

Motor fuel, dispensing operations and equipment
    for hydrogen, 175, 193–198

Motor fuel-dispensing facilities, 174, 175, 189–198
  equipment for hydrogen motor fuel, 175, 193–198
  minimum separation requirements for
    aboveground tanks, 175, 191–192
  warning signs, 175, 189–190

O

Occupancies
  aisle accessways in Group M, 43, 160–161
  change of use or, 1, 2–3

egress from Group A, 43, 170

fire protection for special uses and, 42, 104–109

Group E, 23, 38–39

guard opening limitations for Group R-2, 43, 151–152

minimum door width in Group R-1, 43, 126–127

path of egress travel in Group R-2, 43, 158–159

Occupant load

design, 42, 114–115

determination for day care uses, 42, 116–117

determination for fixed seating, 42, 120–121

Occupant load permitted, maximum, 42, 118–119

Open flame ignition sources in Group R-2 dormitories, 23, 24

Operations, location of spray-finishing, 175, 182–183

Overfill prevention, 203, 244–245

Overfill prevention for LP-gas containers, 203, 267–268

P

Pallets, plastic, 175, 199–200

Panic and fire exit hardware, 43, 134–135

Permit amounts for hazardous materials, 1, 4–5

Petroleum gases, liquefied, 202, 203, 267–273

Physiological warning thresholds, 203, 263–264

Piping systems, control area for fuel in, 203, 209–210

Placards for unsafe buildings, 23, 26–27

Plastic pallets, 175, 199–200

Platform lifts as accessible means of egress, 42, 122

Polyester resin, unsaturated, 203, 252–254

Portable electric space heaters, 41, 44–45

Portable fire extinguishers

electronic monitoring for, 42, 86–87

for solid-fuel cooking appliances, 42, 78–79

Portable tanks, designed, construction, and capacity of, 203, 246–248

Post-fireworks display inspection, 203, 241

Powered industrial trucks and equipment, refueling of, 23, 25

Precautions against fire, general, 23–29

Pressure control systems for refrigeration systems, emergency, 41, 50–53

Prevention, overfill, 203, 244–245

Processes and uses, special, 174–200

flammable finishes, 174, 175, 176–184

high-piled combustible storage, 174, 175, 199–200

motor fuel-dispensing facilities, 174, 175, 189–198

repair garages, 174, 175, 189–198

semiconductor fabrication facilities, 174, 175, 185–188

Projection limits on freestanding objects, 42, 112–113

Protection, weather, 203, 256–257

R

Rack storage, shelves in, 175, 199–200

Ramp length, minimum, 43, 147–148

Ramps

edge protection at, 43, 149–150

handrails for stairways and, 43, 142–144

Records for fire protection systems, 42, 68–69

Reference standards, 274

Refrigerant leak detectors, 41, 46

Refrigeration systems

emergency control boxes for, 41, 48–49

emergency pressure control systems for, 41, 50–53

remote controls for, 41, 47

Refueling of powered industrial trucks and equipment, 23, 25

Refuge

communication systems for areas of, 43, 173

required areas of, 42, 123–125

Remote controls for refrigeration systems, 41, 47

Remote unlocking of stairway doors, 43, 132–133

Repair garages, 174, 175, 189–198

equipment for hydrogen motor fuel, 175, 193–198

minimum separation requirements for aboveground tanks, 175, 191–192

warning signs, 175, 189–190

Residential exterior doors, thresholds at, 43, 130–131

Residential Group R, definition for, 1, 20–21

Resin, unsaturated polyester, 203, 252–254

Roof-hatch openings, protection at, 43, 145–146

S

Safety requirements comment general, 23–39

emergency planning and preparedness, 23, 30–39

general precautions against fire, 23–29

Seating, occupant load determination for fixed, 42, 120–121

Semiconductor fabrication facilities, 174, 175, 185–188

hazardous production materials at a workstation, 175, 185–188

Shelves in rack storage, 175, 199–200

Shutoff controls and valves, definitions for
emergency, 1, 8–9
Signs
exit, 43, 172
warning, 175, 189–190
Site requirements, building and, 41–173
building services and systems, 41
fire protection systems, 41
interior finish, decorative materials, and
furniture, 41
means of egress, 41
Sleeping units, door swings in, 43, 128–129
Smoke and heat vents
for ESFR sprinklers, 42, 97–98
for Group H, 42, 99–100
Smoke venting and draft curtains for high-piled
storage, 42, 101–103
Smokeproof enclosures, egress from, 43, 168–169
Solid-fuel cooking appliances, portable fire
extinguishers for, 42, 78–79
Space heaters, portable electric, 41
Space under stairways, enclosed usable, 43,
140–141
Spaces
egress through adjoining tenant, 43, 156–157
egress through intervening, 43, 153–155
Special processes and uses, 174–200
flammable finishes, 174, 175, 176–184
high-piled combustible storage, 174, 175, 199–200
motor fuel-dispensing facilities, 174, 175,
189–198
repair garages, 174, 175, 189–198
semiconductor fabrication facilities, 174, 175,
185–188
Special uses and occupancies, fire protection for,
42, 104–109
Spray-finishing operations, location of, 175,
182–183
Sprinkler systems, automatic, 42, 72–73, 74–75
Sprinklers, smoke and heat vents for ESFR, 42,
97–98
Stairway doors, remote unlocking of, 43, 132–133
Stairways
curved, 43, 136–137
enclosed usable space under, 43, 140–141
unenclosed interior exit, 43, 166–167
Stairways and ramps, handrails for, 43, 142–144
Standards, reference, 274
Standpipe hose connections at horizontal exits,
Class I, 42, 84–85
Standpipe systems for marinas and boatyards, 42,
82–83

Stationary battery systems, 41, 54–58
Storage
basement, 203, 250–251
of fueled equipment in buildings, 23, 28–29
high-piled, 42, 101–103
LP-gas, 203, 269–273
rack, 175, 199–200
of unsaturated polyester resin, 203, 252–254
Storage cabinets, number of, 203, 249
Storage Group S-1, definition for, 1, 22
Storage warehouses, liquid, 203, 242–243, 255
Swings, door, 43, 128–129
Systems
automatic, sprinkler, 42, 72–73, 74–75
building services and, 41, 44–58
emergency control boxes for refrigeration, 41,
48–49
emergency pressure control systems for
refrigeration, 41, 50–53
emergency voice/alarm communication, 42,
92–93
fire protection, 41, 42, 68–109
monitoring of fire alarm and detection, 42, 95–96
piping, 203, 209–210
records for fire protection, 42, 68–69
remote controls for refrigeration, 41, 47
standpipe, 42, 82–83
stationary battery, 41, 54–58

T

Tanks
aboveground, 175, 191–192
compressed gas, 203, 225–226
fuel, 203, 209–210
portable, 203, 246–248
Tenant spaces, egress through adjoining, 43,
156–157
Testing of hazardous materials equipment, devices,
and systems, 203, 211–213
Thresholds
physiological warning, 203, 263–264
at residential exterior doors, 43, 130–131
Townhouses, automatic sprinkler systems and, 42,
72–73
Toxic gases, treatment systems for, 203, 265–266
Toxic materials, highly toxic and, 202, 203,
263–266
physiological warning thresholds, 203, 263–264
treatment systems for toxic gases, 203, 265–266
Trucks and equipment, refueling of powered
industrial, 23, 25

U

University buildings, Group R-2 college and, 23, 38–39

Unlocking of stairway doors, remote, 43, 132–133

Unsafe buildings, placards for, 26–27

Unsaturated polyester resin, storage of, 203, 252–254

Use or occupancy, change of, 1, 2–3

Uses, special processes and, 174–200
  flammable finishes, 174, 175, 176–184
  high-piled combustible storage, 174, 175, 199–200
  motor fuel-dispensing facilities, 174, 175, 189–198
  repair garages, 174, 175, 189–198
  semiconductor fabrication facilities, 174, 175, 185–188

V

Valves, definitions for emergency shutoff controls and, 1, 8–9

Ventilation termination point, 175, 184

Venting, smoke, 42, 101–103

Vents, smoke and heat, 42, 97–98, 99–100

Violations, missed, 1, 6–7

Voice/alarm communication systems, emergency, 42, 92–93

W

Warehouses, liquid storage, 203, 242–243, 255

Warning signs, 175, 189–190

Weather protection, 203, 256–257
  of exterior egress components, 43, 138–139

Workstation, hazardous production materials at a, 175, 185–188